Alexander Horn
Ultra-fast Material Metrology

Related Titles

Gross, H. (ed.)

Handbook of Optical Systems
6 Volume Set

2010
ISBN 978-3-527-40382-0

Meschede, D.

Optics, Light and Lasers
The Practical Approach to Modern Aspects of Photonics and Laser Physics

2007
ISBN 978-3-527-40628-9

Misawa, H., Juodkazis, S. (eds.)

3D Laser Microfabrication
Principles and Applications

2006
ISBN 978-3-527-31055-5

Saleh, B. E. A., Teich, M. C.

Fundamentals of Photonics

1991
ISBN 978-0-471-83965-1

Diels, J.-C., Rudolph, W.

Ultrashort Laser Pulse Phenomena (2nd Ed.)
Fundamentals, Techniques, and Applications on a Femtosecond Time Scale

2006 Copyright © Elsevier
ISBN 978-0-12-215493-5

Alexander Horn

Ultra-fast Material Metrology

WILEY-VCH

WILEY-VCH Verlag GmbH & Co. KGaA

The Author

PD Dr. Alexander Horn
Kassel University
Institute of Physics
34132 Kassel
Germany

All books published by Wiley-VCH are carefully produced. Nevertheless, authors, editors, and publisher do not warrant the information contained in these books, including this book, to be free of errors. Readers are advised to keep in mind that statements, data, illustrations, procedural details or other items may inadvertently be inaccurate.

Library of Congress Card No.: applied for
British Library Cataloguing-in-Publication Data: A catalogue record for this book is available from the British Library.
Bibliographic information published by the Deutsche Nationalbibliothek
The Deutsche Nationalbibliothek lists this publication in the Deutsche Nationalbibliografie; detailed bibliographic data are available on the Internet at <http://dnb.d-nb.de >.

© 2009 WILEY-VCH Verlag GmbH & Co. KGaA, Weinheim

Printed in the Federal Republic of Germany
Printed on acid-free paper

Typesetting   le-tex publishing services GmbH, Leipzig
Printing   betz-druck GmbH, Darmstadt
Binding   Litges & Dopf GmbH, Heppenheim

ISBN 978-3-527-40887-0

for Kathrin

A Personal Foreword

The continuing trend of the manufacturing technologies towards micro and, most recently, nano applications requires a new generation of production processes as well as integrated process monitoring systems. A spatial resolution of the used production processes in the micro- and nanometer range is obligatory for parts of the same dimension to be manufactured. Due to the specific characteristics such as ultrashort pulse durations in the pico- and femtosecond regime and ultra-large repetition rates in the megahertz regime, the use of ultrashort pulse laser radiation for materials processing is predestined for ablation as well as modification processes in the spatial resolution range to be achieved. For a short time, new high-power femtosecond laser sources with pulse durations in the sub-picosecond regime, repetition rates in the megahertz regime, and mean powers in the multiple hundred watt regime are available. These new laser sources enable a potential economic materials processing of macro parts with femtosecond laser radiation. Ultra-fast, nanoscale processes such as ablation, and modification can be applied to macroscale part dimensions and are almost material independent. Conventional, melt dominant processes such as drilling and cutting can be improved on the basis of sublimation processing, requiring less systems engineering, helping saving resources, achieving higher machining qualities, and higher process qualities yielding in higher productivities: Nano goes macro! The in-hand disquisition on "Ultra-fast Material Metrology" by my respected colleague, and former associate, Alexander Horn, contributes significantly to the more and more deeply investigated, more widely applicable, and increasingly more relevant materials processing with femtosecond laser radiation as well as to the diagnostics needed for process monitoring and control. Yours sincerely,

Aachen, March 2009   *Prof. Dr. rer. nat. Reinhart Poprawe M.A.*

Ultra-fast Material Metrology. Alexander Horn
Copyright © 2009 WILEY-VCH Verlag GmbH & Co. KGaA, Weinheim
ISBN: 978-3-527-40887-0

Contents

Ultra-fast Material Metrology. Alexander Horn
Copyright © 2009 WILEY-VCH Verlag GmbH & Co. KGaA, Weinheim
ISBN: 978-3-527-40887-0

Preface to the First Edition

Laser technologies using cw and long-pulsed laser are well established in many industrial applications, including cutting, welding, and drilling. Since being invented in the 1970s, ultra-fast lasers have evolved into stable systems which enable new applications and enhance quality or productivity of established production techniques due to the different time-scales during absorption of radiation matter and the subsequent processes, such as heating, melting and evaporation. Now ultra-fast lasers are becoming industrially advanced, reaching kilowatt average powers and fulfilling the requirements of industrial standards, such as sealed and hands-off systems.

This book attempts to provide a bridge from research to engineering and gives an overview of the actual ultra-fast laser technologies for scientists and engineers having basic knowledge of laser technology. In this way the ultra-fast laser radiation is described with focus on its application in production. During interaction of this radiation with matter, many processes are initiated. Applying ultra-fast laser radiation with intensities $> 10^{12}\,\mathrm{W/cm^2}$ results not only in heating, melting, and evaporation, but also the formation of high-energy plasma. Thus, plasma physics is introduced for laser material processing.

Industrial applications using laser radiation are mostly linked to process monitoring and control. Ultra-fast laser radiation induces fast processes, which themselves require ultra-fast diagnostics. These diagnostics are presented in this book. Additionally, because of the unique properties of ultra-fast laser radiation, like broad spectral bandwidth and ultrashort pulse duration, this book establishes new diagnostic techniques. This book gives some examples of non-imaging and imaging techniques.

Kassel, March 2009 *PD Dr. rer. nat. Alexander Horn*

Acknowledgment

This book was developed and written during my work as the chair for laser technology (LLT – Lehrstuhl für Lasertechnik) of the Rheinisch-Westfälische Technische Hochschule Aachen (RWTH Aachen). As group leader of the ultra-fast group, I had the chance to work together with top-class scientists, engineers as well physicists, and so was able to explore the fascinating world of ultra-fast laser technology.

My success and progression, and also this book would not have been possible without Prof. Dr. rer. nat. Reinhart Poprawe M.A., Director of the Fraunhofer Institute for Laser Technology (Fh-ILT) and professor of the LLT. He relied on me all the time and satisfied nearly all my scientific wishes. His brave surgency promoted me during my stay in Aachen, and animated me to write this book, adding decisive ideas.

My ex-boss, Akad. Rat. Dr.-Ing. Ingomar Kelbassa (LLT), was the best scientific sparring partner I had at this time. His confidence in my work was unlimited and so enabled a boundless research by my group and me. His precise view on my work clarified often what I wrote bumbling.

I experienced many attractive understanding, working together with Dr. rer. nat. Peter Russbüldt. Also during holidays we spent together he followed my scientific problems and gave determined suggestions for my research and also for this book.

Dr. rer. nat. Ilja Mingareev worked closely with me in this fascinating world of ultra-fast laser sciences. I believe that we made a really good team and therefore made this book achievable and full of interesting results. I thank him also for his patience with me.

My first boss, Dr. rer. nat. Ernst-Wolfgang Kreutz, I owe my first steps in science as a physicist. His rigorous and precise view on the public image of my work impresses me and I thank him for that. He believed all the time in my abilities.

This book was realized especially by the friendly help of Prof. Dr. Martin Richardson, Professor of Optics in the School of Optics at the University of Central Florida. He supported me especially in the field of plasma physics and found time for me at any moment.

I appreciate very much the support of all the proofreaders, Prof. Dr. rer. nat. Wolfgang Schulz, Dr. rer. nat Peter Russbüldt, Dr. rer. nat. Ilja Mingareev, Dipl.-Ing. Martin Dahmen, Dr. rer. nat. Ernst-Wolfgang Kreutz, Bruce Carnevale and Dipl.-Phys. Dirk Wortmann. They cleaned, sorted and eliminated some bad word-

Ultra-fast Material Metrology. Alexander Horn
Copyright © 2009 WILEY-VCH Verlag GmbH & Co. KGaA, Weinheim
ISBN: 978-3-527-40887-0

ings and wrong statements. All the students who made the basic work, set up the experiments, cleaned up the measured data, and made the research presentable I thank very much. I hope they liked as much I did to work together with them. I hope I was not too strict.

All the colleagues of the chair for laser technology and of the Fraunhofer Institute for Laser technology I thank for the good collaboration, the amicable atmosphere, and the familial environmental during all my stays.

Aachen, March 2009   *PD Dr. rer. nat. Alexander Horn*

1
Introduction

> In our highly complex and ever changing world it is reassuring to
> know that certain physical quantities can be measured and pre-
> dicted with very high precision. Precision measurements have
> always appealed to me as one of the most beautiful aspects of
> physics. With better measuring tools, one can look where no one
> has looked before. More then once, seemingly minute differences
> between measurement and theory have led to major advances in
> fundamental knowledge. The birth of modern science itself is inti-
> mately linked to the art of accurate measurements.
>
> T.W. Hänsch. Nobel Price lecture 2005 [1]

1.1
Motivation

The demand for custom-made products with higher machining precision has always pushed the development of new methods. The down-sizing of technologies
into the micro- and nanometer scale prompts the need for a new generation of
processing tools. Ultra-fast laser radiation can fulfill many of the corresponding
requirements. This new generation of radiation processing tools demands the development of focusing methods, beam delivering and new process handling and
surveillance equipment. The complete chain from the laser source as a tool to the
work piece has to be designed from the viewpoint of micro-nanoscaling.

The unique properties of ultra-fast laser radiation enable new technologies for
engineering due to its ability to deposit optical energy in precisely localized volumes and as a no-contact tool. The deposition of optical energy occurs without
inducing stress into the material. The necessity for ultra-fine machining and manipulation with high throughput has moved scientists to develop new laser sources
with key properties like ultrashort pulse duration, ultra-high repetition rates, or
both simultaneously. High-repetition rate systems like fiber, slab or disk-lasers for
high throughput are expected to become industrially applicable soon.

The involved processes themselves like melting, evaporation, and plasma-
formation, can be in general quite complicated due to the simultaneous presence

Ultra-fast Material Metrology. Alexander Horn
Copyright © 2009 WILEY-VCH Verlag GmbH & Co. KGaA, Weinheim
ISBN: 978-3-527-40887-0

of multiple phenomena. Most of the phenomena are on an ultra-fast time scale, demanding also new diagnostic techniques for imaging. Many of these processes are still little understood and have to be investigated for a deeper understanding. Here ultra-fast optical metrology using femtosecond laser radiation is the key technology needed to enable the observation, detection and highlighting of the processes on time-scales $\ll$ 1 ns, or measurement of process variables during micro- and nanostructuring.

For the transfer from the laboratory into the production the new ultra-fast application methods, with the processes involved during ultra-fast machining, have to be controlled adopting ultra-fast optical metrology implying the following:

1. Understanding of the laser-induced processes like melting, evaporation or plasma-formation in advance
2. Detectability and manipulability of process parameters during machining.

In production technology today ultra-fast laser radiation is not widely adopted, because complex processes are expected. By using ultra-fast laser radiation additional processes are given to those in laser engineering with conventional laser radiation. But many of these processes are negligible; some examples are:

- Time-scales for absorption of the ultra-fast laser radiation in the dense matter and in an evolving plasma-plume are separated, in contrast to nanosecond laser radiation. The interaction of an ultra-fast laser pulse with the generated plasma can be omitted.
- Time-scale for absorption is much smaller than the relaxation time of matter for phase change from solid to vapor. This implies that after absorption of ultra-fast laser radiation, matter is excited independent of the material properties.
- Matter is instantaneously[1] excited into a plasma state due to the high intensities of the focused laser radiation. The new machining tool "ultra-fast laser radiation" enables nearly melt-free ablation by the ultrashort pulse duration in the femtosecond regime.
- Typical limitations of machining by wavelength-dependent absorption of laser radiation is mostly overruled by multi-photon processes using ultra-fast laser radiation. Ultra-hard matter, like diamond or tungsten carbide, are difficult to machine using conventional milling techniques, whereas using ultra-fast laser radiation enables machining of usually non-machinable matter. The scales of the created features using femtosecond laser radiation are beyond the resolution of conventional techniques as is well-shown for laser drilling of metals in Figure 1.1.

Ultra-fast laser radiation as an operative tool has advanced now to an engineering level and, as a consequence, ultra-fast engineering technology is even advanced

[1] transition time $< 10^{-14}$ s

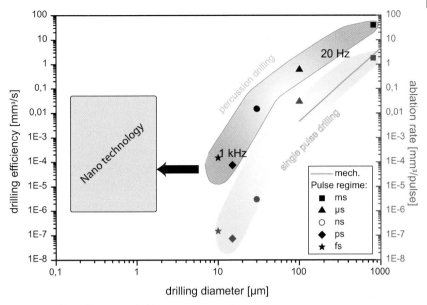

Fig. 1.1 Drilling efficiency and ablation rate versus diameter for mechanical and laser drilling by single pulse and percussion drilling with different pulse durations (data from [2]).

enough to become a new field within the mechanical engineering sciences. To transfer this technology to industry, the new field of ultra-fast optical metrology presented here will document the induced processes by ultra-fast laser radiation and the technologies necessary to detect these processes in order to control them. The physics of ultra-fast laser radiation-matter interaction with the processes of absorption, polarization, evaporation and ionization will be systematically ordered, elucidating key properties of ultra-fast laser radiation adopted in ultra-fast optical metrology.

Ultra-fast optical metrology using pump and probe techniques, utilizing laser radiation simultaneously to initiate and to detect a process, opens the detection with temporal resolutions up to 10 fs and spatial resolutions of 20 nm and below. The methodology and the applications of this new field of research is presented: the choice and the characterization of the probe radiation are fundamental issues for a successful measurement.

The transfer to high throughput ultra-fast engineering is feasible with the enhanced understanding of the involved processes gained by ultra-fast optical metrology. The processing can be controlled by new technologies for ultra-precise manipulation using ultra-fast optical metrology. As a consequence, ultra-fast optical metrology today is ready for applications in ultra-fast engineering technology.

1.2
Definition of Optical Pump and Probe

Optical pump and probe technique is a measurement technique using laser radiation. The laser radiation exhibits at least two functions:
1. Pumping by processing matter, exhibiting, for example, excitation, melt, ablation, evaporation and ionization,
2. Probing to monitor these processes.

In the case of optical pump and probe techniques the radiation beams are extracted from one laser source or from two laser sources. Using one laser source, the radiation is divided into at least two beams, and one beam is delayed temporally to the other in order to monitor using the investigated process (Figure 1.2). The temporal resolution is given by the pulse duration of the laser radiation.

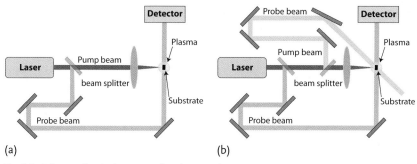

(a)   (b)

Fig. 1.2 Scheme of optical pump and probe metrology using one probe (a) and two probe beams (b).

Ultra-fast radiation adopted in a pump and probe set-up enables one to investigate laser-induced micro- and nanostructuring with > 10 as temporal and 20 nm spatial resolution due to its high temporal resolution, on the one hand, and on the other hand due to the unique properties of this radiation-like broad spectral distribution and small coherence.

1.3
Guideline

The following chapters will give an insight into the optical pump and probe metrology for ultra-fast engineering. The working tool – ultra-fast laser radiation – is described by its market position (Section 1.4), and deals with a resume on the history of high-speed metrology before the invention of the laser, where in the nineteenth century pump and probe metrology became popular. Following this, the history of laser ultra-fast metrology will be sketched.

In the second chapter an overview of the ultra-fast laser source will be given. The sources (Section 2.1), the properties of the focused laser radiation (Section 2.2), and the tools to manipulate and move the beam spatially (Section 2.3), are described concluding with some outstanding challenges in ultra-fast metrology (Section 2.4), and especially the optical pump and probe techniques (Section 2.5).

The third chapter is dedicated to some fundamentals of laser radiation-matter interaction. The interaction of laser radiation at intensities $I > 10^{12}$ W/cm^2 will be discussed, also non-linear processes (Section 3.1), like laser-induced multi-photon absorption (Section 3.2). The plasma generated at these intensities gets a dominant position in the laser-induced process, due to its extreme energetic properties (Section 3.3).

The ultra-fast laser radiation is a fundamental tool of optical pump and probe metrology and will be discussed under this point of view in the fourth chapter. The properties of ultra-fast laser radiation and its propagation through matter, which is important for pump and probe metrology, will be described (Section 4.1). The conditioning of the laser radiation is the scope of Section 4.2; in the case of coherent processes, the generation of quantum states is meant. Probe radiation is often used as an illumination tool, and for nanotechnology applications the imaging is conditional to the resolution limit of the optical system (Section 4.3). The pump and probe technique enables one to detect at different time-steps a process by delaying the probe radiation relative to the exciting pump radiation. Methods and limits of temporal delaying are described in Section 4.4.

The fifth chapter describes the methodology of optical pump and probe in practice, showing also limits of this metrology. The chapter is subdivided into non-imaging and imaging detection. A selection of non-imaging detection methods like spectroscopy (Section 5.1) will be given, furthermore a selection of imaging detection methods describing some actual used set-ups for imaging techniques (Section 5.2) will be presented.

A selection of applications for optical pump and probe metrology for engineering is given in the sixth chapter for drilling and structuring metals (Sections 6.1 and 6.2) and for marking and welding glasses (Sections 6.3 and 6.4).

The seventh chapter describes the perspectives of this new field of research and forecasts optical pump and probe metrology for the future, showing potential applications by using new laser sources (Section 7.1) and also new detectors in combination with improved pump and probe methods (Section 7.2).

1.4
Matrix of Laser Effects and Applications

Ultra-fast laser applications are derived from given applications. As proposed by Sucha [3] and shown in Figure 1.3, the properties of ultrashort laser radiation, such as "high-speed", "high-Power", "bandwidth", "structured spectral coherence" and "short coherence length", drive different phenomena, like ablation, THz imaging,

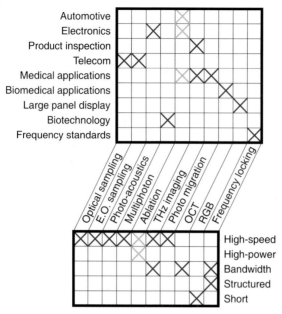

Fig. 1.3 Matrix of laser properties, techniques and applications (according to [3]).

OCT (Optical Coherence Tomography) and frequency conversion for the RGB laser (Red-Green-Blue). From these phenomena the markets are derived. The resulting routes from the properties to the phenomena to the market are many-fold, demonstrating that ultra-fast technology is becoming increasingly market-present in different areas.

From this matrix and the markets it can be deduced that ultra-fast engineering – mainly depicted by the phenomena "ablation" – is connected by two properties of ultra-fast laser radiation: small pulse duration and large peak power. Ablation includes cutting, joining and milling, being the main market for applications with classical laser sources, like cw-CO_2, and cw-Nd:YAG lasers. For ultra-fast mechanical engineering, the property "ablation" emerges with nearly melt-free ablation. This enables new approaches for micro- and nanotechnology applications.

1.5
Historical Survey of Optical Ultra-fast Metrology

1.5.1
Metrology Techniques Before the Advent of Laser

Time-resolved detection of optical emission in former times was mainly limited by the mechanical shutter restricting the temporal resolution to about 1 ms. High-speed metrology has been used in photography since the nineteenth century by using flash-bulbs having microsecond time resolution. High-speed photography was developed in 1834 using a mechanical streak camera. Mechanical streak cameras use a rotating mirror or moving slit system to deflect the light beam. They are limited in their maximum scan speed by mechanical properties, and thus the temporal resolution is limited to about 1 µs [4]. The first practical application of high-speed photography was Eedweard Muybridge's investigation on whether horses' feet were actually all off the ground at once during a trot. Muybridge had successfully photographed a horse in fast motion using a series of twenty-four cameras.

Schlieren photography was developed in 1864 by adapting Foucault's knife-edge test for telescope mirrors to the analysis of fluid flow and propagating shock waves [5]. In 1867 August Töpler combined this set-up using a light spark with about 1 µs emission duration and was able to detect sound waves in air.

A type of Kerr-shutter was invented in 1899 called the Abraham–Lemoine shutter with temporal resolutions of about 10 ns [6, 7]. Two polarizing filters were mounted at 90° to block all incoming light. A Kerr-cell, which changes the polarization of the passing radiation when energized, was placed between the filters and used as a shutter, energized for a very short time by, for example, a spark, and allowed a detector like a photographic plate connected to an imaging system or a spectrometer to be properly exposed.

In 1930 stroboscopes were used for the first time to study synchronous motors [8–10]. Also the dynamics of high-velocity particles like bullets were measured in 1960. Temporal resolutions up to 100 ns could be achieved by flashing. The rapatronic camera, developed in 1940, utilized two polarizing filters and a Kerr-cell to overcome the mechanical limitation of a camera's shutter speed with shutter times of about 10 ns and has been adopted for photography of ab initio nuclear experiments [11, 12].

The improvement of electronics in 1950 enabled the development of electro-optical streak cameras based on a evacuated tube with a photon-sensitive cathode emitting electrons after irradiation and an electron-detecting phosphor screen [13, 14]. The electrons are accelerated in an electric field toward the positively charged phosphor screen. Temporally resolved information with > 200 fs resolution is given by streaking spatially the electrons orthogonally to their propagation direction applying a second, step-like high-voltage electric field. The streaked electron beam induces an optical emission at the phosphor screen, which can be detected by conventional photography.

1.5.2
Ultra-fast Pump and Probe Metrology

The advent of the laser in 1960 [15] created many new fields of research in physics, new applications in mechanical engineering, and pushed optical metrology. Using dye lasers, invented in 1964 [16], a broad spectrum of wavelengths becomes accessible for time-resolved spectroscopy. The dye laser has been brought into the ultra-fast picosecond regime by introducing mode-locking [17]. Ultra-precise spectroscopy is achieved less by the short pulse duration than by small spectral line widths of the laser sources being ideally in the cw-mode (and consequently not described in this book).

Ultra-fast time-resolved measurements are well established in physics, chemistry, and physical chemistry, where fundamental time-scales of selected chemical reactions become accessible for predicting the kinetics of chemical reactions. The photosynthesis of hemoglobin (Hb) has been investigated by transient absorption spectroscopy which could be described as a fast photo dissociation for HbO_2 and a slow dissociation for HbCO [18]. Temporal shaping of the ultra-fast laser radiation enabled chemically induced reactions and control by cooling the vibration modes of HBr molecules [19, 20]. The dissociation of NaI into its constituents has been measured and modeled [21]; in solid-state physics and electrical engineering, carrier dynamics and transport have been probed on picosecond time scales, being directly relevant to the operation of modern high-speed devices [22].

Also, ultra-fast investigations adopting optical pump and probe metrology on an atomic scale and attosecond time scale have been performed using ultra-fast laser radiation. For example, high harmonics X-ray attosecond probe radiation has been used on a Kr atom, which was ionized in the inner shell, generating a hole. The hole recombines with an electron from the outer shell emitting an additional electron, called an Auger-electron. A second laser pulse probes the Kr atom after generation of the electron hole. A life time of 8 fs for the electron-hole has been measured [23–25].

Many investigations have been made by investigating laser-induced processes in condensed matter. Electronic units made from semiconductors like gallium arsenide have a dominant position for information and communication technology. The miniaturization of these elements is not finished today. The processes to generate the circuits on these materials are becoming more and more difficult because the resolution limit for processing techniques using UV radiation is today limited to 60 nm. The detection of the processes during the generation of features with dimensions < 50 nm improves understanding [26]. Pump and probe photo-emission spectroscopy of metals like Au and Ta, has been used to detect the electron dynamics of solids at the surface [27, 28]. For crystalline silicon the thermalization times after irradiation with ultra-fast laser radiation have been measured for the electrons to about 100 fs and respectively for the phonons to about 50 ps [29]. Also, after excitation of GaAs with femtosecond laser radiation using spectral broad probe radiation, the complex dielectric function has been investigated on measuring the reflectivity. The complex dielectric function changes in time from semiconductor to metallic behavior [30].

Imaging using laser radiation as an illumination source has been done since the invention of the laser itself. Some of the latest results are given here. Shadowgraphy of ablation plumes during percussion drilling of metals has been achieved, demonstrating the complex expansion and dynamics of plasma [31] and of the vapor by using fluorescent emission photography [32]. The laser-induced modifications in glasses have been detected by time-resolved interferometry [33] and holography, as well as speckle interferometry, have been adopted using solid state, gas and excimer lasers for different scientific fields.

The experiments described were mostly achieved in the focus of scientific research. In order to emphasize ultra-fast metrology for mechanical engineering, substantial definitions and solution for ultra-fast metrology are given in the following chapters.

2
Ultra-fast Engineering Working Tools

Ultra-fast optical metrology for mechanical engineering uses pulsed laser sources with pulse durations < 10 ps in micro- and nanotechnology, like drilling or structuring, because of the negligible thermally induced load of the irradiated matter and the larger reproducibility of the generated parts.

Ultra-fast metrology for mechanical engineering is described in the following explaining the principle function of femtosecond laser sources necessary for processing and for diagnostics (Section 2.1). For micro- and nanotechnology applications, the ultra-fast laser radiation has to be focused. Femtosecond laser radiation exhibits a group velocity v_g and a phase velocity v_p like pulsed radiation with pulse durations $t_p > 1$ ps. Contrary to pulsed radiation with pulse duration $t > 1$ ps where $v_p = v_g$, due to the broad spectral distribution of femtosecond laser radiation, both velocities have to be considered in addition (Section 2.2). The laser radiation generated by these sources has to be guided to focusing units and the focused beam has to be moved relatively to the substrate in order to process it (Section 2.3). The challenges given by using femtosecond laser radiation for metrology is given by describing the temporal, spatial and spectral domains (Section 2.4). Optical pump and probe techniques are analyzed through the described domains (Section 2.5).

2.1
Ultra-fast Laser Sources

Ultra-fast lasers adopted in micro- and nanotechnology are pulsed lasers with pulse durations $t_p < 10$ ps. Some general aspects on those sources are given in Section 2.1.1. Also, general aspects on ultra-fast laser sources used in optical pump and probe metrology are given in Section 2.1.2. Ultra-fast lasers used in micro- and nanotechnology consist of a laser source and, for applications with larger pulse energies, an ulterior amplifier. Femtosecond laser oscillator principles are given in Section 2.1.3. Femtosecond laser radiation is amplified by chirped-pulse amplification (CPA) and will be described in Section 2.1.4. Facilities, even not being reasonable for industrial applications, exhibit radiation with extraordinary properties for micro and nano-processing as well for pump and probe metrology (Section 2.1.5).

Ultra-fast Material Metrology. Alexander Horn
Copyright © 2009 WILEY-VCH Verlag GmbH & Co. KGaA, Weinheim
ISBN: 978-3-527-40887-0

The knowledge gained using radiation from facilities can be implemented into industrial applications.

2.1.1
General Aspects for Ultra-precise Engineering

Ultra-precise engineering needs tools with ultra-high precision and high productivity. Using ultra-fast laser radiation enables ultra-fine application of optical energy and ultra-high precision by coupling optical energy into matter on an ultra-fast timescale. Furthermore it enables large productivity by ultra-high repetition rate of the laser radiation.

- **Dosage:** An exact measurement can be achieved due to the highly defined threshold for ablation or modification working with laser radiation at pulse durations < 10 ps. The optical energy can be adopted to change the properties of matter (Chapter 3). For example, material can be reproducibly, deterministically, and non-stochastically ablated with intensities 0.1–0.5% above the ablation threshold. With ablation rates below 10 nm/pulse and geometry widths ≈ 1µm nano-structuring enables the generation of intelligent topologies like "Lotus effect" surfaces.
- **Precision:** Ablation and modification by ultra-fast laser radiation is nearly independent from the physical and chemical constitution of matter due to the ultra-fast coupling of optical energy. For example, a negligible amount of heat is transferred to the surrounding of a metal[2] resulting in instantaneous phase transitions from solid into the vapor state (Section 3.3.2). For multi-photon processes, the beam diameter in the focus can be reduced well below the diffraction limited Gaussian beam diameter $w_{\text{eff}} \approx \lambda/n$, n multi-photon factor (Section 2.2). Combined with the ultra-fine application of optical energy, ultra-precise structuring for mechanical engineering is enabled.
- **Productivity:** Repetition rate is the key parameter of the ultra-fast laser radiation for productivity: for compensation of the small ablation rates – needed for ultra-precision – a laser source with large repetition rates in the MHz-regime is necessary. For large productivity at high precision using large repetition rate radiation, an ultra-precise and very fast positioning system is also needed (Section 2.3). For the precision of the workpiece positioning with respect to the laser focus, also the temporal and spatial beam properties of the laser source, like beam point stability influenced by deflection of the beam by process gas induced turbulence have to be considered (Section 2.2.5).
- **Industrial requirements:** Long-term stability, low-maintenance, simple operation and being affordable are main necessities of an industrial application of ultra-fast lasers. The stability and durability of laser sources has been improved in recent years since laser diodes have been implemented as pumping sources and taking benefit from fiber technology, developed for telecommunication and consumer products. Different laser concepts are

---

2) heat affected zone (HAZ) < 100 nm

now available: "classical" amplified diode-pumped solid-state laser systems are becoming ready solutions and fulfill more and more the needs for an industrial environment. Ultra-fast solid-state lasers based on rod crystal geometries historically use the older technology, meaning Ti:sapphire laser systems being pumped by a frequency-converted q-switch Nd:YLF laser. Alternatively ultra-fast fiber lasers are becoming maintenance-free due to the "simple" design and the mounting, as well as alignment-free set-up. Due to the evolving material science for new laser materials, fiber lasers are also becoming applicable with pulse energies above 1 µJ at large repetition rates. The reliability increase results in a reduction of the costs when transitioning this technology to production. Today commercially available ultra-fast lasers are limited to pulse energies < 100 µJ.

2.1.2
General Aspects on Ultra-fast Lasers for Ultra-fast Metrology

Ultra-fast metrology needs a tool with manifold properties: Laser radiation with ultra-fast pulse durations is needed for the photography of fast evolving processes; super-continuum spectra (bandwidth) have to be generated and analyzed for spectroscopic applications. The unique properties of ultra-fast laser radiation for metrology can be highlighted by:

- **Temporal resolution:** The most obvious property of ultra-fast laser radiation is the high temporal resolution. During, for example, sawing, drilling, milling most processes in "classical" mechanical engineering like heating, melting, vaporization and plasma formation are "frozen in" as detected by some kind of indicator, like photography using ultra-fast pulse durations below 1 ps. The technique of pump and probe has been greatly improved, by exchanging the probing source from flash-bulb with pulse durations of about 1 µs for pulsed laser radiation with pulse durations as small as 5 fs.
- **Peak power:** Multi-photon processes are initiated and become significant for laser radiation with peak power $P_p >$ MW focused to spot sizes in the µm-range. Peak powers of this value, considering that the peak power defined as $P_p \approx E_p/t_p$, are easily generated by laser radiation with pulse duration of less than 100 fs and a pulse energy of 100 nJ. Focusing laser radiation with such peak power delivers in-focus intensities between 1 TW/cm^2 and 1 PW/cm^2. The electric fields associated with such intensities can ionize atoms with field ionization and multi-photon ionization. Further acceleration of these electrons can induce avalanche ionization in ulterior atoms and subsequently generate additional ions and electrons.
- **Spectrum:** The Heisenberg uncertainty principle for energy and time

$$\Delta E \cdot \Delta t \geq \hbar, \tag{2.1}$$

can be translated to $\Delta E \cdot t_p \geq 1$ for photons. This principle means that for ultra-fast laser radiation, broad spectra result due to the high temporal certainty given by the pulse duration t_p. For example, the spectral bandwidth of laser radiation with a pulse duration of 100 fs at a central wavelength of 800 nm is about 20 nm. Reducing the pulse duration to 5 fs the bandwidth increases > 100 nm.

- **Spectral coherence:** One property of ultra-fast laser radiation is the inherently coherent broad spectrum of the laser radiation. In a time-averaging spectrometer a spectral "comb" with a peak separation represents the repetition rate of the laser radiation. Frequency measurement is one leading area in which the coherence properties of this frequency comb are of primary importance and will be put into products quickly.
- **Temporal coherence:** The temporal coherence of ultra-fast laser radiation is correlated with the pulse duration. The coherence length, defined by $L_C = \lambda^2/\Delta\lambda$ with the spectral bandwidth $\Delta\lambda$, is related to bandwidth transform limited laser radiation to the pulse duration t_p by

$$L_C \approx ct_p . \tag{2.2}$$

- **Phase coherence:** The state in which two signals maintain a fixed phase relationship with each other or with a third signal that can serve as a reference for each is a coherent phase. Laser radiation with stabilized carrier envelope offset (CEO), exhibits a fixed phase. The time dependence of the electric field associated with an optical pulse can be described as a fast sinusoidal oscillation, called the carrier, multiplied with a more slowly varying envelope function. [3]

For deeper insight to the fundamentals, design and development of such sources, information see [3, 34, 35]. Generally an ultra-fast laser source consists of at least one laser oscillator and optionally one or more amplifiers. The typical repetition rate of a ultra-fast laser oscillator for micro- and nano-processing is in the range 10–100 MHz. Therefore, a laser oscillator with an average power of 0.1–1 W produces pulsed laser radiation with energies of approximately 1–10 nJ. This pulse energy is sufficient for metrology of, for example, thin films by using picosecond laser-induced acoustics waves, for experiments that probe heat transfer or carrier dynamics, and for many methods of optical spectroscopy. These pulse energies are not generally large enough for materials modification, except with laser radiation that is tightly focused by high-numerical-aperture microscope objectives. Larger pulse energies are available from "extended-cavity oscillators" that operate with a lower repetition rate, on the order of 10 MHz, but this technology is currently limited to pulse energies < 100 nJ in commercial Ti:sapphire lasers and < 1 μJ in Yb:tungstate lasers. For the machining of materials, the laser radiation from laser oscillators must be amplified to reach energies > 1 μJ [36].

3) http://www.rp-photonics.com/

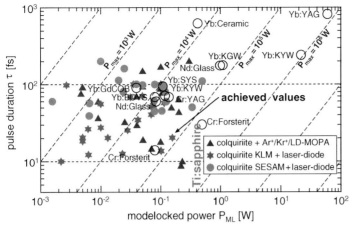

Fig. 2.1 Distribution of ultra-fast solid-state oscillators with repetition rates about 100 MHz for pulse duration, peak power, and mode-locked power [35].

2.1.3
Laser Oscillator

Radiation with pulse durations < 10 ps is generated by mode-locked laser oscillators. The phases of the longitudinal optical modes of the laser radiation in the laser cavity are locked together by self-induced processes, also called passive effects, such as Kerr-lensing (KLM) in the gain medium, by additive pulse mode-locking (APM), or by use of a saturable absorber. ultrashort pulsed laser radiation with a high repetition rate is produced by mode-locking, determined by the length of the optical cavity.

The typical repetition rates for these solid-state oscillators are about 100 MHz. For pulse durations larger than 1 ps direct diode-pumped solid-state laser systems with active media based on rod, fiber or disk geometry are in use. Typical media are Nd:YAG, Nd:glass, Nd:YVO$_4$, Yb:glass, Yb:YAG, and Cr:LiSGAF (Figure 2.1 and Table 2.1).

2.1.3.1 Rod and Disk Solid-State Laser

Femtosecond pulses are generated by passive mode-locking (with a saturable absorber), basically because a saturable absorber, driven by already short pulses, can modulate the resonator losses much faster than an electronic modulator: the shorter the pulse gets, the faster the obtained loss modulation, as long as the absorber has a sufficiently small recovery time. The pulse finally reaches as pulse duration the recovery time of the absorber of ≈ 100 fs. In some cases, reliable self-starting mode-locking is not achieved.

The scientific market is still dominated by Ti:sapphire lasers adopting passive mode-locking being actually a set of three lasers (Figure 2.2a). Continuous-

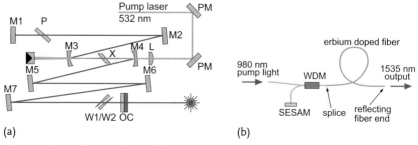

Fig. 2.2 Scheme of a Ti:sapphire solid-state oscillator (a), and of an erbium-doped mode-locked fiber laser (b).

wave (cw) diode lasers pump a cw solid-state laser, which is frequency-doubled ($\lambda = 532\,\text{nm}$) and used to pump the Ti:sapphire oscillator (Figure 2.2). The cavity is designed for a form of passive mode-locking called soft aperture Kerr-lens mode-locking, supporting a stable pulse operation without a so-called hard aperture [37]. The Ti:sapphire crystal is pumped at $\lambda = 532\,\text{nm}$ with $\approx 5\,\text{W}$ pump power. For laser cavities delivering ultrashort pulses a folded cavity around the crystal is built up using dispersion compensating mirrors M1–M7 [38]. The fine-tuning for dispersion can be achieved by additional BaF$_2$ substrates P and W[4]. The wide bandwidth of optical radiation over 100 nm for sapphire doped with titanium enables small pulse durations well below 4 fs [39].

Today, passively mode-locked ultra-fast oscillators are used for chronometry, because very precise round-trip times are achieved by stabilization of the resonator length and cavity envelope offset (CEO). Super-broad white-light continua generated by focusing ultra-fast laser radiation into a dielectric material, for example, silica fibers or photonic fiber (Chapter 3.1). The frequency combs are generated by superimposing many pulses and are used as a ruler for radiation with unknown frequency: one of the known frequencies in the comb will result in the unknown frequency into a detectable beat mode. This beat mode is easy to measure using a simple photo diode [1].

2.1.3.2 Fiber Oscillator

The advantages in ultra-fast laser technology, such as precision frequency metrology [40], absolute optical phase control using CEO [41, 42] and generation of coherent frequency comb [43, 44], today can be achieved by ultra-fast fiber laser oscillators (Table 2.1).

Passively mode-locked oscillators operate without modulators using components that act on the amplitude of the oscillating pulses, for example, SESAM[5] (Figure 2.2)[3]. Additive-pulse mode-locking [45, 46] is a technique for passive mode-locking adopted, for example, for fiber lasers, used for generating short optical

---

4) http://www.iqo.uni-hannover.de/morgner/ tisa.html

5) Semiconductor saturable absorber mirror made in semiconductor technology. Such

a device contains a Bragg mirror and (near the surface) a single quantum well absorber layer

Table 2.1 Commercial laser oscillators based on rod and fiber crystal geometries.

	Manufacturer	Model	Central wavelength (nm)	Repetition rate (MHz)	Pulse duration (fs)	Average power (mW)
Rod	Amplitude	t-Pulse 500	1030	10	< 500	5000
	Coherent	Chameleon Ultra	850	80	140	2500
	Coherent	Micra	800	78	100	300
	Coherent	Vitesse 800	800	80	100	750
	Del Mar	Mavericks	1250	76	65	< 250
	Del Mar	Trestles Series	820	83	>20	< 2500
	Femto Lasers	FemtoSource20S	800	80	20	900
	Femto Lasers	Synergy	800	75	10	400
	High-Q	FemtoTRAIN Nd	1060	72	200	100
	High-Q	FemtoTRAIN Ti	800	73	100	200
	KMLabs	Cascade	795	< 80	> 15	100
	KMLabs	Graffin	790	90	> 12	450
	MenloSystems	Octavius	800	1000	6	330
	Newport	MaiTai HP	800	80	100	< 2500
	Newport	Tsunami	800	80	100	< 2700
Fiber	Clark MXR	Magellan	1030	37	200	40
	Del Mar	Buccaneer	1560	70	150	100
	IMRA	ULTRA AX-20	780	50	100	20
	IMRA	ULTRA BX-60	1560	50	100	60
	MenloSystems	C-Fiber	1560	100	100	250

pulses with femtoseconds pulse durations. The general principle of additive-pulse mode-locking is to obtain an artificial saturable absorber by exploiting nonlinear phase shift in an optical single-mode fiber.

For applications in instrumentation and as seed sources for high-power amplifiers, passively mode-locked fiber lasers are the systems for applications which do not necessitate larger power and good pulse duration-bandwidth product (PBP) [3]. Laser radiation from an ultra-fast fiber laser with larger pulse energies can be achieved by using passively mode-locked multi-mode fiber lasers. Fiber oscillators operated at 1.56 μm can be frequency-doubled to produce laser radiation with a wavelength of around 800 nm, which is an emission wavelength for many currently considered applications of ultra-fast engineering.

2.1.4
Amplifier

2.1.4.1 Amplification Media

The ability to amplify ultra-fast laser radiation has been an elusive goal for 20 years following the development of the ultra-fast oscillator in the mid-1960s. In 1985, high-intensity ultra-fast lasers emerged with the development of the chirped-pulse amplifier (CPA) [47].

Twenty years later, Ti:sapphire chirped-pulse amplifiers producing 1–2 mJ optical pulses at 1 kHz repetition rates are available from several commercial suppliers (Table 2.2). Higher-repetition-rate (> 100 kHz) lasers with some μJ pulse energy are desirable for many applications in materials removal and modification. The relative simplicity of amplifiers that are directly pumped by diode lasers is driving the development of systems based on Nd:YAG, Nd:glass, Er:glass, and Yb:WO$_4$. These systems may lead to more compact and less expensive instruments, inhibiting a more limited range of output wavelengths and a larger pulse duration than Ti:sapphire [36]. Table-top ultra-fast laser systems emit laser radiation with pulse energies < 1 nJ applied in metrology. Today laser system are available with laser radiation at pulse energies > 1 J adopted for relativistic optics phenomenon called plasma wake-field acceleration of electrons that may one day replace synchrotrons as bright sources of X-rays [48–50]. For pulse energies ≫ 1 J commercial systems based on Ti:sapphire CPA technology are available with pulse energies E_p < 100 J. Laser facilities with multi-beam lines have been set-up delivering up to E_p < 1 kJ ultra-fast laser radiation at very small repetition rates of f_p ≪ 0.05 Hz.

2.1.4.2 Chirped-Pulse Amplification (CPA)

In a chirped-pulse amplifier, pairs of diffraction gratings are used to temporally stretch the optical pulse prior to amplification and then temporally compress the pulse after amplification (Figure 2.3)[6]. All these laser sources are based on a Master-Oscillator-Power-Amplifier (MOPA) principle: laser radiation with small pulse energies in the range 1–10 nJ is generated by an ultra-fast laser oscillator, called a seeder, working at large repetition rates in the range 10–100 MHz. From this laser radiation one pulse is extracted at much smaller repetition rates. Depending on the final pulse energy the repetition rate varies between 10^4 Hz at pulse energies < 1 mJ and 10^{-4} Hz at pulse energies ≫ 1 kJ. The radiation of the seeder has been amplified in at least one or several amplifier systems. The main difference between a MOPA laser system with pulse durations > 100 ps and ultra-fast laser systems is found in the large peak intensities attainable by ultra-fast laser radiation, being above threshold intensities of optical material. Before amplification the peak intensity has to be reduced by enlarging the pulse duration of the laser radiation to many hundred picoseconds in order to avoid damage of the amplifier optics. Enlarging of the pulse duration is achieved by chirping the pulse using diffraction

---

6) http://www.eecs.umich.edu/USL/
       HERCULES/index/index.html

Table 2.2 Commercial laser amplifiers based on rod and fiber crystal geometries.

	Manufacturer	Model	Central wave-length (nm)	Repetition rate (kHz)	Pulse duration (fs)	Average power (W)
Rod	Amplitude	s-Pulse	1030	10	400	1
	Clark-MXR	CPA 2210	775	2	150	2
	Coherent	Legend Elite fs	800	1–3	130	1
	Coherent	Legend Elite ps	800	1–5	500–2000	3
	Coherent	Legend Elite ultra	800	1–5	35	3
	Coherent	Legend HE Cryo	800	1–5	50	25
	Continuum	Terawatt	800	0.01	100	2
	Femto Lasers	FemtoPower	800	1, 3	10, 30	1
	High-Q	femtoREGEN	1035	1–40	300–600	1
	High-Q	picoREGEN	532, 1064	0–100	12000	10
	KMLabs	Dragon	780	1–10	30	30
	KMLabs	Wyvern	800	50–200	50	2
	Newport	Solstice	800	1, 5	100	2
	Newport	Spitfire Pro	800	1, 5	35, 120, 2000	5
	Quantronix	Cyro Amplifier	800	1	30–120	12
	Quantronix	Integra-C	800	1	40	3
	Quantronix	Integra-HE	800	1	40	7
	Quantronix	Odin II	800	1	30	3
	Thales	Alpha 1000s	800	1–10	100	5
	Thales	FemtoCube	785	1–10	30–100	2.5
	Thales	Bright	785	1–5	120	1.5
Fiber	Clark-MXR	Impulse	1030	0.2–25	250	20
	Fianium	Femtopower1060 XS	1064	40	300	5
	IMRA	μJewel1000	1045	0.2–5	350	1.5
	Polaronyx	Uranus 3000	1030	0.01–0.05	500	1
	Polaronyx	Uranus 2000	1030	0.01–0.1	600	1

gratings, prisms or dispersive fibers. After amplification the pulse duration is reduced again using grating or prism compressors. Large intensities are affordable by using laser radiation with large pulse energies and short pulse durations. At intensities above 10^{12} W/cm^2 the amplified laser radiation has to be carried through evacuated beam lines.

2.1.4.3 Optical Parametric Chirped-Pulse Amplification (OPCPA)

The concept of chirped-pulse amplification was originally developed for the amplification of ultrashort pulses with laser amplifiers, but it was soon realized that it

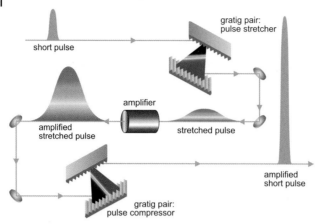

Fig. 2.3 Principle of chirped pulse amplification (CPA): stretching and compression of ultra-fast laser radiation by gratings and amplification.

is also very suitable for optical parametric amplifiers (OPAs). At high pulse energies, these also profit from a strong reduction of the peak intensities by amplifying temporally stretched (chirped) pulses.

Compared with classical chirped-pulse amplification based on laser gain media, OPCPA has a number of important advantages:

- The parametric gain within a single pass through a nonlinear crystal can be many tens of db, so that OPCPA systems require fewer amplification stages (often just one), usually do not involve complicated multi-pass geometries, and can thus be built with much more simple and compact setups.
- Parametric amplification is possible in a wide range of wavelengths.
- With optimized phase-matching conditions, the gain bandwidth can be very large, allowing few-femtosecond high-energy pulses to be generated.

Although using highly nonlinear quasi-phase-matched crystals enables very high gains with moderate pump pulse energies, they generate radiation with pulses durations > 100 fs and energies in the range µJ–mJ. Such systems can be made very compact, cost-effective and efficient.[7]

2.1.4.4 Amplifier Designs
Depending on the amplification stage, the amplifier systems used are designed in different ways:

- regenerative amplification by temporal superposition of the laser radiation,
- multi-pass amplification by spatial superposition of the laser radiation, and
- single-pass amplification by fiber or Innoslab amplifier.

7) http://www.rp-photonics.com/optical_
parametric_chirped_pulse_amplification.
html

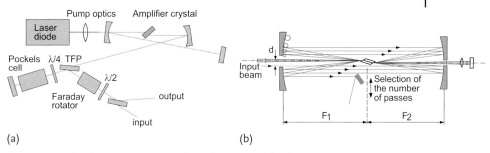

Fig. 2.4 Principle of a regenerative (a) and a multi-pass amplifier (b) [51].

Regenerative amplifiers are achieved by placing the gain medium in an optical resonator, in combination with an optical switch, usually realized with an electro-optic modulator and a polarizer (Figure 2.4a[8]). The laser radiation passes multiple times through the gain medium during amplification. The optical switch injects the pulses into the amplifier before amplification and couples it out afterward. As the number of round trips in the resonator can be controlled by an optical switch, it can be very large, so that a very high overall amplification factor is achieved by clearing all the saved energy in the active material within many round trips. The amplification is controlled in the range 10^5 to 10^6 by the number of round trips of the laser pulse to be amplified. Only intra-cavity pulse energies up to 10 mJ are amplified due to the small damage thresholds of the Pockels cells and Faraday isolators.

Multi-pass amplifier systems are constructed using unstable resonator geometries. The laser radiation from the seed-source passes the laser medium many times, and the laser radiation is amplified stepwise at each passage by a factor of 3–10 (Figure 2.4b). The number of passes is limited by the geometry of the design and by the increased difficulty of focusing all the passes in the pumped volume of the medium. The number of passes is usually limited to 4–8. For higher gains, several multi-pass amplifiers can be cascaded. To earn gain the crystal is pumped by frequency-converted Q-switch laser radiation. Amplification of large pulse energies and peak power is realized by single-pass amplifier systems. In order to reduce the damage probability the beam line is evacuated and the beam diameter is enlarged up to 50 cm. Electro-optical elements like Pockels-cell crystals, thin-film polarizers, Faraday rotators and lenses induce non-linear dispersion, like the group velocity dispersion (GVD) and the third order dispersion (TOD) (Section 3.1.1). After amplification the pulse duration of the laser radiation has to be compressed. Depending on the spectral width of the ultra-fast laser radiation, the compression is achieved using gratings or prisms. Small spectral widths, for example, $\Delta\lambda \approx 25$ nm follows $\Delta\omega \approx 11$ THz at $\lambda = 800$ nm, and a minimum pulse duration of

$$\Delta t = \frac{2\pi K}{\Delta\omega} \approx 166 \text{ fs} ,\qquad(2.3)$$

8) http://www.rp-photonics.com/regenerative_
amplifiers.html

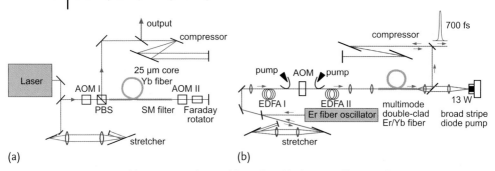

Fig. 2.5 Double-pass (a) and 3-amplifier Y-doped high-energy fiber CPA (b) [3].

with $K \approx 0.3$ for sech^2-intensity distribution [52]. The GVD of optical materials can be compensated by grating or prism compressors. Prism compressors are used for the low-loss compensation of additional TOD. Optical materials usually have a positive TOD which can not be compensated by grating compressors.

Fiber amplifiers use as a laser-active dopants Ytterbium and Erbium. High-power large energy ultra-fast laser radiation is generated due to the recent emergence of Yb as a dopant providing a more suitable medium than Er. The main advantages of Yb-doped fibers are:

- broader amplification band (50–100 nm compared to 10–30 nm), and
- larger optical pumping efficiency (60–80% compared to 30–40%).

An all-fiber-based Yb-doped CPA emitting laser radiation with 100 μJ pulse energy and about 220 fs pulse duration has been demonstrated (references in [3]). The principal set-up consists of an oscillator, an external pulse stretcher, three amplifier stages and an external pulse compressor (Figure 2.5a) [3]. Output energies of up to $E_p = 1$ mJ are obtained from a slightly more sophisticated setup (Figure 2.5b). The set-up consists of a fiber-based oscillator, a diffraction grating stretcher, a three-stage Yb fiber amplifier chain with two optical gates between the stages and a diffraction grating compressor. At $f_p = 1667$ Hz repetition rate the uncompressed laser radiation exhibits $E_p = 1.2$ mJ pulse energy and after compression a pulse duration $t_p = 400$ fs [3].

The advent of high power cw-fiber lasers in the production lines of automotive industries for the cutting and welding of steel, for example, was followed by the industrial development of pulsed fiber lasers. New technologies for fiber lasers have to be developed due to the non-linear processes induced by laser radiation with large peak-power in the optical elements of the laser.

For example, a 50 W sub-picosecond fiber chirped pulse amplification system generating 50 μJ pulses at a repetition rate of 1 MHz has been demonstrated [53]. As required for precise high speed micro-machining, this system has a practical system configuration enabled by the fiber stretcher and 1780 l/mm dielectric diffraction grating compressor and is capable of ablation rates > 0.17 mm^3/s for metals, ceramics, or glasses.

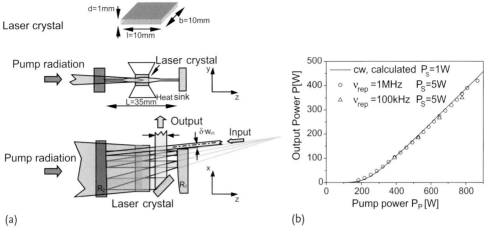

Fig. 2.6 Principle of beam propagation in an Innoslab amplifier (a) and output power as function of pump power and repetition rate (b) [56].

A promising alternative to a fiber amplifier is the Innoslab amplifier developed at the Fraunhofer Institute for Laser Technology (Fh-ILT) [54]. An Innoslab amplifier consists of a longitudinally placed, partially laser diode-pumped slab crystal [55]. The short distance between the pumped gain volume and the large cooled mounting surfaces allow for efficient heat removal. The line-shaped pumped cross section inside the slab matches the beam characteristic of laser diode bars. Innoslab lasers are designed for moderate gain of about a factor 2–10 per pass. Nine passes through the slab crystal with an amplification factor of 1000 are achieved [56, 57] (Figure 2.6) with a confocal cavity. Nevertheless, Innoslabs are single-pass amplifiers. At each passage a new section of the gain volume is saturated. In Innoslab amplifiers beam expansion on every passage through the slab balances the increase of power and intensity.

Two concepts applying an Innoslab amplifier have been demonstrated amplifying ultra-fast laser radiation at quasi-cw operation with 76 MHz repetition rate and at pulsed operation with 0.1–1 MHz repetition rate:

- Chirped pulse amplification (CPA) of ultra-fast laser radiation is an established technology to avoid damage of the optics and distortion by nonlinear effects of the radiation. To attain high stretching factors free space dielectric gratings setups have to be used. These gratings can handle very high average power > 8 kW, but scaling the average power into the 100 W regime is a considerable additional operating expense. An Innoslab amplifier seeded with ultra-fast laser radiation with repetition rates > 10 MHz needs no CPA technology [56] (Figure 2.6). Because of the small amount of material in the optical path, nearly transform-limited ultra-fast laser radiation is achieved with 400 W average power and 682 fs pulse duration. To achieve an average output power > 300 W a pump power of $P_{pump} = 800$ W have to be deliv-

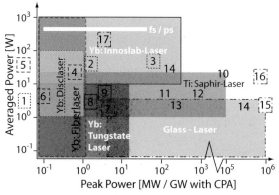

Fig. 2.7 Available high-power ultra-fast lasers: 1. Jenoptik,
2. Time-Bandwidth, 3. Trumpf, 4. Corelase, 5. IAP FSU Jena,
6. PolarOnyx, 7. Univ. of Michigan, 8. Amplitude Systems,
9. Light Conversion, 10. APRC Japan, 11. Coherent, 12. MBI
Berlin, 13. Spectra-Physic, 14. Thales 100/30, 15. NOVA (LLNL)
Phelix (GSI) Vulcan (RAP), 16. POLARIS (FSU Jena), 17. Innoslab (Fh-ILT) [58].

ered by two horizontal laser diode stacks consisting of four collimated laser
diode bars each. The pump radiation is imaged in a planar waveguide for
homogenization in the slow axis and finally imaged by a relay optic into the
laser slab crystal.

- Using a similar setup to the previous Innoslab amplifier, but seeding with
ultra-fast laser radiation exhibiting a smaller repetition rate 0.1–1 MHz, CPA
technology has been applied resulting in 420 W average power ultra-fast
laser radiation with 720 fs pulse duration at 0.1–1 MHz repetition rate [57]
(Figure 2.6b). Applying Innoslab amplifiers the average power frontier has
been extended into the kW regime (Figure 2.7).

2.1.4.5 Commercial Systems

Laser systems with large pulse energies and low repetition rates are mostly de-
ployed for academic use and based on solid-state rod geometry (Figure 2.7). Large
pulse energies in the range from 1 mJ to 100 J are the common ultra-fast laser
systems based mainly on Ti:sapphire, Nd:glass, Yb:glass, or Cr:LiSAF laser media
(Table 2.2). These laser systems are capable of generating laser radiation with pulse
durations as small as 5 fs. Pump and probe experiments, using ultra-fast probe
radiation with extraordinary properties (Section 2.1.2) are applied, for example,
to time-resolved investigations on laser-induced melting of crystals by femtosecond X-ray reflectometry [59], or time-resolved atomic dynamics by attosecond X-ray
spectroscopy [23, 24, 60].

Laser systems capable of being implemented into industrial production lines are fully sealed picosecond or femtosecond laser systems based on direct diode pumped regenerative amplified systems. These systems work with crystals like Yb:KGW for the generation of femtosecond laser radiation. Ti:Sapphire laser systems are also becoming more and more stable in long-term operations. They have the drawback of necessitating a pumping source in the green spectral region, today solved by frequency-doubled pulsed or cw Nd:YLF/Nd:YAG or $Yb:VO_4$ lasers. This increase in complexity is also an additional source of failure (Table 2.2).

Large repetition rates are necessary for large productivities, but are not readily available at large pulse energies today. Ultra-fast processes are generally activated and investigated in small interaction volumes. Machining of metals for the industrial demand, for example, can be handled using ultra-fast laser radiation at small pulse energies $\ll 1$ mJ. Increasing the repetition rates above 100 kHz requires a redesign of the laser system. Limitations to the frequency are given by Pockels-cells, enabling a maximum response frequency of some hundreds of kHz. Acousto-optic modulators in the amplifier stages are one approach to overcome this. The other approach is to modulate the seed laser radiation reducing the repetition rate from some 100 MHz to some 100 kHz with acousto-optic devices before amplification in a multi-pass or fiber amplifier.

In particular, industrial laser systems have to be fully sealed to become insensitive to environmental influence. Disks as active medium generating laser radiation with < 1 ps pulse duration and 50 W average power at 40–60 MHz repetition rates are an attractive alternative to rods. The thermal budget is fully controlled due to the very thin disk, resulting in high-quality ultra-fast laser radiation, being scalable without thermal distortions of the radiation.[9]

2.1.5
Facilities

Common laser systems today are Ti:Sapphire laser systems for the moderate pulse energy regime up to 5 J and Nd:glass laser systems for the high-end systems up to 1.8 MJ pulse energies. Peak powers over 1 PW are generated, for example, by the Megajoule laser of LIL or the Petawatt laser of LLNL (Figure 2.7). These last mentioned laser systems are single shot lasers used for fundamental research.

Large pulse energies up to kJ and (in the future, MJ) are used (mainly) in fundamental research[10]. For the industrial realization of micro- or nano-processing these systems are not practical. Indeed, some special research topics can only be carried out by these facilities due to the wide range of available wavelengths in the range 1 nm–10 μm and intensities $< 10^{22}$ W/cm^2 achieved by these facilities. More details are given in the Addendum D.

9) http://www.timebandwidth.com/
10) http://www.answers.com/topic/
inertial-confinement-fusion

2.2
Focusing of Ultra-fast Laser Radiation

Engineers and researchers deal increasingly with laser radiation and optical systems. The knowledge of the propagation of laser radiation through optical systems is essential. An important impact on the success of the applications using ultra-fast laser radiation has the precise definition of the radiation parameters [61].

Focusing laser radiation needs optical elements. The requirements for ultra-small foci are described in Section 2.2.1. The laser radiation emitted by industrial ultra-fast laser sources can be approximated by Gaussian radiation and is given in Section 2.2.2. The beam parameters for the technical description of laser radiation are given in Section 2.2.3, where the Gaussian beam description enables the introduction of the beam quality, which is an essential parameter for focusing. Unlike "classical" pulsed radiation, where the velocities of the pulse and the phase front are equal, when ultra-fast laser radiation propagates through media the velocities of the phase and the pulse front have to be considered, Section 2.2.4. For strong focusing, the scalar wave equation for the fields and the polarization are no longer valid and have to be replaced by a vectorial description. The precision of processing is determined by the focus diameter and the beam positioning stability. The letter one is described in Section 2.2.5. The key parameters enabling ultra-fast laser radiation to be adopted in metrology are given in Section 2.2.6.

2.2.1
Ultra-small Laser Focus

Focusing of ultra-fast laser radiation is achieved with optical elements like lenses, objectives, mirrors, and combinations of them. The optical laws responsible for focusing laser radiation are described in the field of linear and non-linear optics [62].

Laser radiation can be focused with refractive, diffractive, and reflective objectives. Reflective objectives show no chromatic aberrations. On the other hand, reflective objectives need large design efforts because laser radiation has to pass beside the non-transmissive mirrors, for example, Schwarzschild objectives, using a focusing main mirror and a secondary mirror. An alternative solution is the use of off-axis paraboloids, with the disadvantage of large focal lengths $f > 100$ mm.

Optics describes the propagation of light and of laser radiation. Often for simplicity only the paraxial propagation of unpolarized and spatially circular Gaussian intensity distribution is described [61–63]. The propagation of radiation is limited to cw radiation and pulsed low-intensity radiation. It is described by monochromatic linear optics, whereas ultra-fast laser radiation has to be described by fully-including polychromatic linear optics with all aberrations. Additionally, the temporal and spatial pulse shape, the distribution and the velocity of the pulse front as well of the phase front have to be described [64–66]. Ultra-small focus $\leq 1\,\mu\text{m}$ denotes a vectorial description of the electromagnetic fields and its polarization [67].

Non-linear optics is not related to laws governing the focusing of laser radiation by an objective, but to the matter reacting non-linearly on large intensity radiation.

Matter is, in this case, on the one hand the environment, like air or process gas and on the other hand the substrate under investigation. More details are described in Section 3.1.

2.2.2
Gaussian Beam

Since Gaussian radiation is described within the frame of the paraxial approximation (called Gaussian optics), the description of the propagation with a Gaussian beam fails when wavefronts are tilted by more than about $30°$ from the propagation direction [68]. The description of propagation of radiation by Gaussian beams is valid only for laser radiation with beam waists larger than about $2\lambda/\pi$. In the case of non-Gaussian optics, numerical methods have to be adopted for the calculation of the beam propagation.

Electromagnetic radiation emitted by commercial femtosecond lasers can be approximated by Gaussian radiation, whose transverse electric field and intensity distributions are described by Gaussian functions. Many lasers emit radiation with a spatial Gaussian profile, called the "TEM$_{00}$ mode" or higher-order solutions of the paraxial wave equation, so-called Gauss–Hermite or Gauss–Laguerre modes. When refracted by a lens, radiation with a Gaussian profile is transformed into radiation with a different Gaussian profile.

The Gaussian function is a solution to the paraxial form of the scalar Helmholtz equation in SVE approximation, which represents the complex amplitude of the electric field. For Gaussian radiation the complex electric field amplitude, measured in V/m, at a distance r from its center, and a distance z from its waist, is given by

$$E(r,z) = E_0 \frac{w_0}{w(z)} \exp\left(\frac{-r^2}{w^2(z)}\right) \exp\left(-ikz - ik\frac{r^2}{2R(z)} + i\zeta(z)\right), \tag{2.4}$$

where $k = 2\pi/\lambda$ is the wave number (in radians per meter). The beam radius $w(z)$, the radius of curvature of the wavefronts $R(z)$, and the Guy phase $\zeta(z)$ are described in Section 2.2.3 in more detail. The corresponding time-averaged intensity distribution is

$$I(r,z) = \frac{|E(r,z)|^2}{2\eta} = I_0 \left(\frac{w_0}{w(z)}\right)^2 \exp\left(\frac{-2r^2}{w^2(z)}\right). \tag{2.5}$$

E_0 and I_0 are, respectively, the electric field amplitude and intensity at the center of the beam at its waist, that is $E_0 = |E(0,0)|$ and $I_0 = I(0,0)$. The constant

$$\eta = \sqrt{\frac{j\omega\mu}{\sigma + j\omega\varepsilon}} \tag{2.6}$$

is the characteristic impedance of the medium in which the beam is propagating and in vacuum $\eta = \sqrt{\frac{\mu_0}{\varepsilon_0}}$.

Generally, the focus radius of a Gaussian beam, also called the beam waist w_0, is proportional to the wavelength λ and inversely proportional to the numerical aperture (NA) of an optical system, or by using focal length f and beam radius in front of the objective w_L

$$w_0 \propto \frac{\lambda}{NA} \tag{2.7}$$

$$\propto \frac{\lambda f}{w_L} . \tag{2.8}$$

In order to induce non-linear processes in matter the intensity of the radiation has to be $> 10^{10}\,\mathrm{W/cm^2}$. To ultra-precise material processing, the adopted optical energy should be small, implying ultra-small laser foci in order to reach the required intensities. An ultra-small focus results in an ultra-small focal volume

$$V_F \approx \frac{z_R \pi w_0^2}{4} = \frac{\pi^2 w_0^4}{4\lambda} , \tag{2.9}$$

with

$$z_R = \frac{\pi w_0^2}{\lambda} , \tag{2.10}$$

the Rayleigh length (Figure 2.8). Laser radiation has to be collimated, formed and conducted via mirrors to a focusing system. In contrast to conventional radiation, ultra-fast laser radiation has to be addressed with more attention compared to monochromatic laser sources. Because of the broad spectrum of ultra-fast laser radiation, single optical elements, like dielectric mirrors and lenses, which are corrected for laser radiation at only one wavelength with small bandwidth, cannot be used at all, because chromatic aberration is induced. Nonlinear interaction with the optical material takes place when ultrashort pulsed laser radiation with large peak powers is adopted within the objective. The properties of the optics change, like the transmittance or the spatial refractive index distribution, and finally the optical material degrades. As a consequence, laser radiation is absorbed in optical elements, or, because of self-focusing, is catastrophically focused inside the optical element. Generally, GVD and TOD have to be considered when ultra-fast laser is focused (Sections 2.2.3 and 3.1.1).

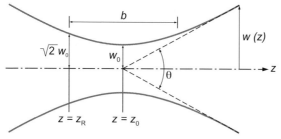

Fig. 2.8 Parameters of a Gaussian beam.

2.2.3
Beam Parameters of Gaussian Radiation

The Gaussian radiation is described by the beam parameters (Figure 2.8):
- beam radius w, and
- Rayleigh length z_R.

These parameters are defined in the following. Laser radiation with a Gaussian intensity profile propagating in free space is described by the beam parameter, the beam radius (spot size) $w(z)$ having its minimal value at w_0. For radiation with the wavelength λ at a distance z from the beam waist, the beam radius is given by

$$w(z) = w_0 \sqrt{1 + \left(\frac{z}{z_R} \right)^2} , \tag{2.11}$$

with the origin of the z-axis at the beam waist w_0, and where z_R is the Rayleigh length given by Eq. (2.10). The beam radius or spot size of the beam, is the radius at which the field amplitude and intensity drop to $1/e$ and $1/e^2$, respectively. At the distance z_R from the waist, the radius w of the beam is

$$w(\pm z_R) = w_0 \sqrt{2} \equiv w_R . \tag{2.12}$$

The confocal parameter or depth of focus of the Gaussian beam is defined as

$$b = 2z_R = \frac{2\pi w_0^2}{\lambda} . \tag{2.13}$$

The radius of curvature of the wavefronts comprising the beam is calculated as

$$R(z) = z \left[1 + \left(\frac{z_R}{z} \right)^2 \right] . \tag{2.14}$$

The beam radius $w(z)$ increases linearly with increasing z for $z \gg z_R$. The angle θ between the beam's central axis and the slope at $z \gg z_R$ is called the far-field divergence of the radiation and for an ideal Gaussian beam the beam parameter product (BPP) is given by

$$\theta \simeq \frac{\lambda}{\pi w_0} \quad (\theta \text{ in radians}) . \tag{2.15}$$

The total angular spread of the Gaussian beam far from the waist is given by $\Theta = 2\theta$.

The quality of laser radiation can be described by a quality factor M^2, which describes the deviation of the nearly diffraction-limited real laser from the ideal diffraction-limited beam, defined as the Gaussian beam. The laser radiation quality M^2 can be described by the ratio of the beam parameter product (BPP) of the technical radiation (θ_{real}, W_0^{real}) and an ideal Gaussian beam (θ_{Gauss}, w_0^{Gauss}):

$$M^2 = \frac{\theta_{real} W_0^{real}}{\theta_{Gauss} w_0^{Gauss}} \tag{2.16}$$

The BPP of technical radiation is obtained by measuring the minimum beam radius and far-field divergence and then taking their product. All real laser beams have M^2 values greater than one, whereas ideal Gaussian beams have $M^2 = 1$. Micro- and nano-structuring claim small beam waists and, in other words, laser radiation with M^2 nearly one.

A more precise description of the beam quality for technical radiation can be given by the variances [69]. In the scalar theory of electromagnetic fields, the electromagnetic field of a monochromatic wave propagating in the z-direction can be written

$$E(x, y, z, t) = \text{Re}\left(\tilde{E}(x, y, z)e^{i(\omega t - kz)}\right) . \tag{2.17}$$

The complex amplitude $\tilde{E}$ can be described by the Fourier transform of the x- and y-coordinates [70], called space-frequency distribution

$$\tilde{E}(x, y, z) = \int_{-\infty}^{\infty} \int_{-\infty}^{\infty} \tilde{P}(s_x, s_y, z) \exp\left[-i2\pi(s_x x + s_y y)\right] ds_x ds_y . \tag{2.18}$$

The space-frequency distribution is the inverse transform of $\tilde{E}$

$$\tilde{P}(s_x, s_y, z) = \int_{-\infty}^{\infty} \int_{-\infty}^{\infty} \tilde{E}(x, y, z) \exp\left[i2\pi(s_x x + s_y y)\right] dx dy . \tag{2.19}$$

The intensity distribution is given by

$$I(x, y, z) = \left|\tilde{E}(x, y, z)\right|^2 , \tag{2.20}$$

$$\cdot \; I(s_x, s_y, z) = \left|\tilde{E}(s_x, s_y, z)\right|^2 . \tag{2.21}$$

In [70] it is described that $I(s_x, s_y, z)$ is independent of z. Every space-frequency component s_i can be described for small angles Θ_i by plane waves propagating with an angle Θ_i to the z-axis

$$s_i = \frac{\sin \Theta_i}{\lambda} \approx \frac{\Theta_i}{\lambda} . \tag{2.22}$$

The Gaussian intensity distribution of the TEM$_{00}$ remains Gaussian, even after a partial Fourier transformation of the x- and y-coordinates (Eq. (2.19)). In this case a space-frequency intensity distribution $\hat{I}(s_x, s_y)$ in the space-frequency domain is given by

$$\hat{I}(s_x, s_y) = I_0 e^{-2\pi^2 w_0^2 (s_x^2 + s_y^2)} . \tag{2.23}$$

The standard deviation of a Gaussian beam in the space and in space-frequency domain is defined by

$$\sigma_x(z) = \sigma_y(z) = \frac{w(z)}{2} , \tag{2.24}$$

$$\sigma_{s_x} = \sigma_{s_y} = \frac{1}{2\pi w_0} . \tag{2.25}$$

The space-bandwidth product (SBP) for diffraction-limited radiation is defined by

$$\sigma_{0,x} \times \sigma_{s_x} = \sigma_{0,y} \times \sigma_{s_y} = \frac{1}{4\pi} . \tag{2.26}$$

Defining a Gaussian diffraction limited multi-mode beam with an intensity distribution $I(x, y, z)$, one can write the variance as

$$\sigma_x^2 = \frac{\int (x - \bar{x})^2 I(x, y, z) dx\, dy}{\int I(x, y, z) dx\, dy} , \tag{2.27}$$

$$\sigma_{s_x}^2 = \frac{\int (s_x - \bar{s}_x)^2 \hat{I}(s_x, s_y) ds_x\, ds_y}{\int \hat{I}(s_x, s_y) ds_x\, ds_y} . \tag{2.28}$$

The caustic is described by

$$\sigma_x^2 = \sigma_{0,x}^2 + \lambda^2 \sigma_{s_x}^2 (z - z_{0,x})^2 \tag{2.29}$$

and results in the SBP for real beams

$$\sigma_{0,x} \times \sigma_{s_x} = \frac{M_x^2}{4\pi} \tag{2.30}$$

with $M_x^2 \geq 1$. It can be shown that $M_{x,y}^2$ is invariant with respect to every non-diffracting paraxial optics [71]. Defining the beam diameter as

$$W_x(z) = 2\sigma_x(z) , \tag{2.31}$$

$$W_{x,0} = W_x(0) = 2\sigma_{x,0} , \tag{2.32}$$

the spatial variance of the beam waist is obtained

$$W_x^2 = W_{x,0}^2 + M_x^4 \frac{\lambda^2}{\pi^2 W_{x,0}^2} (z - z_{0,x})^2 , \tag{2.33}$$

or written with the Rayleigh length of the real beam

$$Z_{R,x} = \frac{\pi W_{0,x}^2}{M_x^2 \lambda} , \tag{2.34}$$

$$W_x^2 = W_{x,0}^2 \left[1 + \left(\frac{z - Z_{0,x}}{Z_{R,x}} \right)^2 \right] . \tag{2.35}$$

With the parameters M_x^2, $W_{x,0}$ and $Z_{0,x}$ the propagation of the radiation through the optical system is fully described. Therefore, a beam diameter of the "embedded" Gaussian beam is defined as [71]

$$2w_{0,x}^{em} = \frac{W_{0,x}}{M_x} . \tag{2.36}$$

The further propagation of the embedded Gaussian beam waist is calculated by adopting the ABCD law and the beam transfer matrices. The position of the waist is not changed by this transformation. The propagation of technical radiation is characterized by three parameters [70]:

1. the beam waist $W_{0,x} = 2\,\sigma_{0,x}$,
2. the beam waist position $Z_{0,x}$, and
3. M_x^2 or the Rayleigh length of real radiation $Z_{R,x}$.

Beam caustics with the beam waist position $Z_{0,x} \neq Z_{0,y}$ are called astigmatic, and for $W_{0,x} \neq W_{0,y}$ asymmetric.

Because of this property, Gaussian radiation focused on a small spot with the radius w_0, exhibits a large divergence Θ. Well-collimated radiation is achieved using the BPP by radiation with small divergence and large beam radius.

2.2.4
Pulse Duration

Focusing ultra-fast laser radiation is achieved by lenses or mirrors and combinations of the two. The ability to achieve large intensities in the focus depends on the ability to keep the pulse duration and the spatial extension of the focus small. A difference between group and phase velocity of the ultra-fast laser radiation in a lens and the group velocity dispersion (GVD) can reduce the intensity in the focus:

- delaying the radiation, passing on the axis of the lens with respect to the radiation arriving from the peripheral of the lens, results in a delay between pulse and phase front,
- displacing the focal points of different spectral components due to the polychromaticity of the laser radiation, and
- changing the pulse duration due to the GVD.

An optical element transforms the phase of an optical plane wave into a spherical one which converges, in the paraxial approximation, to the focus (Figure 2.9). During propagation through the optical material the group velocity v_g is transformed, Eq. (4.4), Section 4.1.2, being different from the phase velocity $v_p = c/n$ by

$$\Delta T(r) = \left(\frac{1}{v_p} - \frac{1}{v_g} \right) L(r), \tag{2.37}$$

with $L(r)$ the thickness of the optical element.

In the case of a spherical lens

$$L(r) = \frac{r_0^2 - r^2}{2} \left(\frac{1}{R_1} - \frac{1}{R_2} \right), \tag{2.38}$$

where $R_{1,2}$ are the radii of curvature of the lens surfaces and r_0 is the radius of the lens aperture. The difference in the travel time from the surface to the focal spot for r_0 and r is given by the chromaticity of the optical element

$$\frac{d}{d\lambda} \frac{1}{f} \tag{2.39}$$

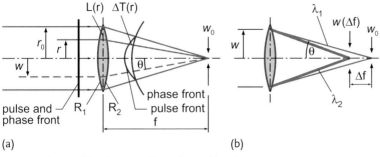

Fig. 2.9 Scheme of the plane wave being focused by a lens (a) and of the chromatic aberration for two wavelength (b) [72].

and for a spherical lens the group velocity delay is given by

$$\Delta T(r_b) = \frac{r_0^2 - r^2}{2} \lambda \frac{d}{d\lambda} \left(\frac{1}{f} \right) \tag{2.40}$$

using $1/f = (n-1)(R_1^{-1} - R_2^{-1})$ for the focal length. For example, ultra-fast laser radiation focused by a lens $f = 30$ mm exhibits a group velocity delay

$$\Delta T(r_b = w_L) = -\frac{w_L^2}{2cf(n-1)} \left(\lambda \frac{dn}{d\lambda} \right) \tag{2.41}$$

resulting in a time difference $\Delta T \approx 300$ fs using the wavelength $\lambda = 248$ nm, the pulse duration $t_p = 50$ fs, and the beam radius $w_L = 2$ mm with the optical specifications $n \approx 1.51$ and $\lambda dn/d\lambda \approx 0.17$. The chromaticity of a lens on the laser radiation, Eq. (2.39), is a spatial spread of the optical energy around the focus

$$\Delta f = -f^2 \frac{d(1/f)}{d\lambda} \Delta \lambda \tag{2.42}$$

$$= -\frac{f\lambda^2}{c(n-1)} \frac{0.441}{t_p} \frac{dn}{d\lambda}, \tag{2.43}$$

for an ultra-fast pulse with a temporal Gaussian pulse shape with the pulse of duration $t_p = \sqrt{2\ln 2}\, t_g$ and a spectral band width $\Delta\lambda = 0.441\lambda^2/ct_p$. For the above-mentioned laser radiation and lens parameters one obtains a spread $\Delta f = 60\,\mu$m. This value is much larger than the Rayleigh length $z_R = w_0/\theta \approx 5\,\mu$m for that lens. The relative widening of the focus can be calculated using Eq. (2.11) to

$$\frac{w(\Delta f)}{w_0} = \sqrt{1 + \left(\frac{\Delta f}{2z_R} \right)^2} \approx \left(\frac{\Delta f}{2z_R} \right) \tag{2.44}$$

$$\overset{(2.43)}{=} -\frac{0.441}{2t_p} \frac{f\lambda^3}{w_0^2 c\pi(n-1)} \frac{dn}{d\lambda}. \tag{2.45}$$

The chromatic broadening is of the same order as the group velocity delay ΔT.

In addition to the temporal group velocity delay and the spectral broadening, a direct temporal broadening by the optical material of the lens is induced through the group velocity dispersion, GVD (Section 4.1.2). The group velocity dispersion effect is pulse duration dependent and becomes significant for NIR-VIS laser radiation with pulse durations $< 100\,\text{fs}$. An increase of the pulse duration by GVD from $50\,\text{fs}$ to about $60\,\text{fs}$ is calculated for unchirped laser radiation with $\lambda = 248\,\text{nm}$, Eq. (3.12), and using the second-order dispersion index of $\lambda d^2 n/d\lambda^2 \approx 2.1\,\mu\text{m}^{-1}$ a lens thickness of $d = 2.1\,\text{mm}$.

A numerical evaluation as described in [72] allows one to study the complex space and time distribution of the intensity between a lens with small numerical aperture and the focus. Because of the properties of polychromatic radiation, an approximation for a radius-dependent pulse delay in the focus was found, Eq. (2.41). An approach to overcome this limitation is the development of an achromatic doublet by Bor [64, 73]. The achromaticity condition

$$\frac{d}{d\lambda}(1/f) = 0 \tag{2.46}$$

gives in addition to the lens equation for a doublet with two refractive indices and ensures full chromatic compensation. Very tight focusing is achieved by commercial microscope objectives with large numerical apertures $0.8 < NA \leq 1.4$, which are only adapted for the wavelength but not for ultra-fast laser radiation properties. Aberrations induced by using femtosecond laser radiation in commercial microscope objectives with larger NA are still investigated today [67]. Methods to detect the pulse duration in the focus are given in Section 4.2.3.

2.2.5
Beam Stability

The stability of laser radiation Θ_s is sometimes an issue treated with neglect because it is difficult to measure. In general, beam stability is understood as the temporal and the spatial stability of laser radiation. For example, the relative beam point stability of solid state laser radiation $\Theta_s/\Theta < 10^{-3}$ is smaller than the relative beam point stability of gas laser radiation, because no turbulence or laser medium in the laser resonator deflect radiation. Fiber lasers have much smaller beam point stability than solid state lasers because the fiber-geometry imposes the allowed radiation mode[11]. But often beam stabilities for long term applications are not specified by laser manufactures.

The spatial intensity distribution is described by temporally averaged values. Similar to the measurement of the beam point stability, no spatial intensity distributions are given, neither for long term values, nor for the pulse to pulse variation. Spatial stabilities as small as 0.1% are reported for ultra-fast fiber lasers.

---

11) Due to thermal lensing the axial position of
    the beam waist is moving in time

The temporal stability of a laser source is represented by the time-dependent pulse energy and pulse profile. The pulse to pulse stability is often quoted as values around 1%, but the energy distribution in one pulse, meaning the background of the ultra-fast pulse described by the nanosecond and picosecond pedestal, is often not stated. The measurement of the pedestals is elaborate, so even commercial solutions are available[12].

Apart from this beam instability from the laser source itself, the laser beam can be deflected and distorted by density changes in the ambient air and by temperature changes in the lab. As a consequence of, for example, density changes of the ambient air the beam position moves and the beam quality varies with the fluctuation period of the turbulence. This can be prevented by tubing the beam and providing laminar flushing gas flow by an inert gas. The temperature-induced geometry changes of mirror holders and other optical parts can be prevented by temperature stabilization, for example, by climatization.

2.2.6
Key Parameters for Ultra-precision Machining and Diagnostics

2.2.6.1 Ultra-precise Machining

For machining of matter by ultra-fast laser radiation, two key parameters can be defined: pulse duration and spatial intensity distribution. Ultra-fast laser radiation can be focused to a very small focus, bearing in mind that due to chromaticity and GVD. Using microscope objectives with $NA > 0.8$, even not optimized for ultra-fast application, very small holes with sub-μm dimensions have been generated by IR-femtosecond laser radiation [74]. The ability to generate smaller geometries by ablation than the optical limit of focused laser radiation allows, can be explained by the spatial Gaussian intensity distribution. By reducing the intensity in the focus very close to the ablation threshold – utilizing the property of very deterministic laser ablation – structuring is achieved adopting only the tip of the Gaussian intensity distribution.

In addition to linear absorption occurring with metals, multi-photon absorption in dielectrics reduces the effective beam waist by the factor $\sqrt{N}$, N being the multi-photon factor. Ablation of matter by multi-photon induced absorption enables sub-μm ablation.

Investigations of the involved processes during irradiation of matter with ultra-fast laser radiation have demonstrated, that ablation behavior changes dramatically if a temporal modulation of the ultra-fast laser radiation is applied [75, 76]. For example, irradiation of dielectrics with temporally asymmetric femtosecond pulses of identical fluence results in different final free electron densities and, as a consequence, different thresholds for surface material modification in sapphire and fused silica [77] (Figure 2.10). Nanostructures have been generated with geometry dimensions below 100 nm.

---

12) http://pagesperso-orange.fr/
amplitude-technologies/sequoia.htm

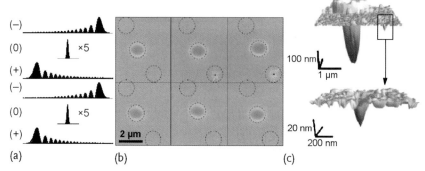

Fig. 2.10 Laser intensities (a), corresponding nanostructures in fused silica detected by SEM (b) and by AFM (c) [77].

By using scanning near-field microscopy (SNOM), the diffraction limit can be overcome and the beam width can be reduced to below 100 nm with the drawback that the intensities are very low. It has been shown that nanostructuring of metals layers is possible with spatial resolution of 50 nm [78].

2.2.6.2 Ultra-fast Pump and Probe Diagnostics for Mechanical Engineering

Key parameters for non-imaging pump and probe diagnostics are the pulse duration and, essentially for time-resolved spectroscopy, the spectral distribution of the radiation. The pulse duration of the probe beam has to be one order of magnitude smaller than the detected process.

For imaging diagnostics a homogeneous and constant spatial intensity distribution is desirable. To achieve this, the mostly Gaussian intensity profile has to be manipulated by beam stops or by diffractive optical elements. The image of an object with dimension close to the applied wavelength is distorted due to the spatial coherence of the radiation. This enforces the reduction of the wavelength into the ultraviolet regime.

2.3
Beam Positioning and Scanning

Micro- and nanostructuring techniques need tools for ultra-fine removal. On the one hand, the laser focus has to be positioned on a target with high-precision on a sub-micrometer scale. On the other hand, due to this very small tool geometry, very large position velocities are needed to achieve large productivity.

To reach *high-precision* a small and reproducible laser focus has to be available, and additionally, the laser beam has to be positioned on a target by positioning stages or by scanning systems. By adopting *positioning stages*, a substrate is moved on mechanical axes, for example, in two dimensions, such as the x- and y-directions, and on a z-axis the laser focus position is changed (Table 2.3). The

Table 2.3 Commercial linear positioning stages.

Model	Axis	Precision μm	Max. velocity mm/s	Max. traverse path mm
PI	Ball screw	1	50	1000
Micos	Ball screw	1	50	200
Kugler	Air	0.01	1000	300
Aerotech	Air	0.05	1000	300
Anorad	Air	0.1	1000	300

overall precision is given by the precision of all positioning stages and by the beam point stability of the laser system (Section 2.2.5).

2.3.1
Positioning

A positioning stage is composed of a movable carriage guided by linear guiding. The precision of a linear guiding is defined by the properties:
- minimum increment motion,
- unidirectional repeatability,
- bidirectional repeatability,
- pitch, and
- yaw.

The overall precision of one positioning axis is determined by the superposition of the properties of linear guidings. Also these properties are a function of the velocity, the acceleration of the axis, and the load. To accomplish a high precision mechanical set-up, bearings have to be manufactured with high-precision and, in order to fulfill rigidity, special alloys and ceramics are used for the guiding and the carriage. The sliding properties of the carriage and of the guide rail are given by the two categories: rolling-element and plane linear bearing (Figure 2.11).

A plane linear bearing is very similar in design to a rolling-element bearing, but it contains no sphere bearings. Typical for simple linear positioning stages is dovetail guiding. In order to accomplish low-friction sliding, different combinations are chosen: bronze, metal/polymer, and all-polymer bushings. The friction is still existent and the precision is modest because of the hard-handle stick-slip phenomena, and so this design will no longer be considered. Air bearings and oil bearings are nearly frictionless concepts, with high precision $\approx 1\,\text{nm}$, uniform motion and no abrasion. This also results in large achievable moving velocities of the carrier. The drawback of pneumatic bearing is the limited load. By using oil instead of air, because of the increased viscosity, the stiffness is increased and the load can be in-

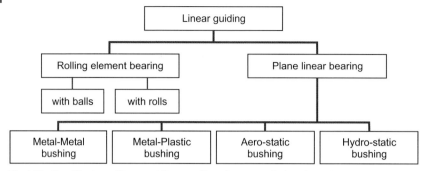

Fig. 2.11 Classification of linear guiding in rolling-element and plane linear bearing [79].

creased but corresponding to increasing friction of the bearings and reducing the maximum velocity of the positioning handling.

A rolling-element bearing is generally composed of rolls or balls. Ball bearings consist of a sleeve-like outer ring and several rows of balls retained by cages. Its features are smooth motion, low clearance, low friction, high rigidity, and long life. Compared with plane bearings, rolling-element bearings are more economical than air bearings but exhibit a precision of only 1 μm. For nanostructuring rolling bearings are not advisable.

For the translation of the carriage three methods have been established:
- Spindle drive with shaft joint uses jack screws to transform a rotation of a stepper motor into a linear movement. A repeatability about 200 nm at velocities up to 1 m/s are precisions achievable today.
- Linear motor drive are directly coupled to the carrier offering direct transmission of a movement of the motor to the carrier with negligible clearance.
- Piezoelectric drive is a type of electric motor based upon the change in shape of a piezoelectric crystals when an electric field is applied. Piezoelectric motors use the converse piezoelectric effect, whereby the material produces acoustic or ultrasonic vibrations in order to produce a linear or rotary motion. The elongation in a single plane is used to make a series of stretches and position holds. Piezo linear motors can be divided into two groups: ultrasonic motors, also referred as resonant motors, and step motors. Resonant motors are characterized by a simpler design and higher speeds, step motors can achieve much higher resolution, larger forces and highly dynamic performance over small distances. A great advantage of a piezoelectric drive is their intrinsic steady-state auto-locking capability. Step and continuous piezo motors are, in principle, nonmagnetic and vacuum-compatible, which is a requirement for many applications in the semiconductor industry.

Positioning detection is essential for micro- and nano-structuring. Using stepper motors the rotation angle per step is known. Calculating the gear transmission

Table 2.4 Commercial scanning systems (calculated for a f-theta objective with focal length $f = 160$ mm).

Model	Model	Aperture in mm	Velocity max m/s	Repeatability μm	Field size mm × mm
GSI	HB X10	10	4.5	10	100 × 100
GSI	HSM15M2	15	2	2	95 × 95
GSI	HPM10VM2	10	5	4	120 × 120
SCANLAB	hurrySCAN 30	30	4.5	2	50 × 50
SCANLAB	intelliSCAN 10	10	15	5	150 × 150

ratio, the linear displacement can be calculated. Alternatively, the linear displacement can be measured. Counting the increments results also in the knowledge of the actual position. Propagation of mechanical imperfections results in a limited resolution of about 2 μm. Higher resolution can be achieved by using a linear encoder. A linear encoder is a sensor, transducer or read head paired with a scale that encodes position. The sensor reads the scale in order to convert the encoded position into an analog or digital signal, which can then be decoded into position by a digital readout (DRO). Motion can be determined by change in position over time. Linear encoder technologies include capacitive, inductive, eddy current, magnetic, and optical encoder. The position resolution achievable by linear encoder is in the range 1–10 nm.

2.3.2
Scanning Systems

Contrary to moving the substrate or the focusing optics, the laser radiation can be positioned by galvano scanning systems (Table 2.4). The laser radiation is deflected by two orthogonally positioned movable mirrors. Galvanometer scanners are high-performance rotary motors for optical applications and consist of a motor section based on moving magnet or coil technology and a high-precision position detector. The motor section of each axis is ideally matched to the inertial load presented by the mirror. The optimized rotor design is responsible for the dynamic properties and resonance characteristics of the mirror. Axially pre-loaded precision ball bearings guarantee a backlash-free rotor assembly with high stiffness and low friction. The optical position detector system is characterized by high resolution, as well as good repeatability and drift values. The scanners are equipped with heaters and temperature sensors. This allows temperature stabilization for further enhancing long-term stability, even under fluctuating ambient conditions.[13]

13) www.scanlab.de/de/

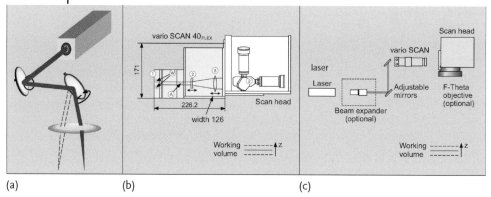

(a)   (b)   (c)

Fig. 2.12 Principle of a galvanometer scanner (a), focusing before the scanner unit (b), and combination of focusing before and after scanning unit (c)[13].

The first mirror deflects the laser radiation in the x-direction and the second mirror deflects the laser radiation in the y-direction (Figure 2.12). The mirrors are moved by closed loop galvanometer-based optical scanners[14]. Fundamentally, galvano scanners have operating swing frequencies at the resonance frequency of the mirrors. In order to sustain large positioning velocities, the resonance frequency of the galvanometer scanner has to be large. The mass scales with the mirror area. Increasing the area to more than $20 \times 20 \, \text{mm}^2$ increases the moment of inertia by increasing the mass and decreasing the resonance frequency

$$\omega \propto \sqrt{\frac{1}{L}} = \sqrt{\frac{1}{mv^2}}.$$

(2.47)

To accomplish this the mirrors have to be lightweight.

Three designs are used for scanning systems, deflecting the radiation before or after the focusing objective:

- The first design, deflecting the laser radiation and afterward focusing the laser radiation with an objective, is adopted for large positioning precision and small focused beam diameters using small focal lengths (Figure 2.12a)[15]. The scanning area ranges from $100 \, \mu\text{m}^2$ to some $10\text{--}100 \, \text{cm}^2$. F-theta lenses have been optimized for focusing laser radiation in a plane field using a galvanometer scanning system. Deflected from this point, laser radiation can be imaged perpendicular to the image field. For short pulse laser scan optics, special low dispersion glass has to be used in order to minimize the broadening of a laser pulse caused by the group velocity dispersion.[16]

14) www.cambridgetechnology.com
15) http://www.thefabricator.com/LaserWelding/
    LaserWelding_Article.cfm?ID=1278
16) http://www.silloptics.de/english/products/
    produkte-1/laser.html

- The second design focuses the laser radiation by a lens in front of the scanning system, and afterward the laser radiation is deflected by the galvanometer scanner (Figure 2.12b[17]). Due to the fact that the laser radiation is convergent, the focal length of the focusing unit is large ($f \gg 100$ mm), the beam diameter at the mirror is limited by the ablation-threshold fluence and, consequently, the focal beam diameter is limited. Additionally, the position of the laser radiation in propagation direction can be controlled by moving coaxially the focusing objective. The scanning regime ranges from 100 mm^2 to some m^2.
- The third one is a combination of the first two described designs enabling radiation with smaller focal lengths. During the scanning process, a movable beam expander is positioned along the optical axis with respect to a stationary focusing optic (Figure 2.12c) and produces a change in the systems overall focal length, synchronized with the mirror motion. The movement of the laser focus in the propagation direction is controlled by a first optical system before the galvanometer scanner[18], and a second optical system behind the galvanometer scanner finally focuses the radiation. Focusing in a volume is achieved enabling, for example, microablation of hard metals with complex structures[19] with scanning areas from 1 mm^2 to some 100 mm^2 and scanning velocities up to 10 m/s (at $f = 160$ mm). The scanning velocity depends on the focal length.

2.4
New Challenges to Ultra-fast Metrology

Production technologies introducing ultra-fast laser sources require new techniques for metrology. Technologies, like semiconductor electronics for micro- and nanotechnology as well as genomics for life-science, need tools for process monitoring and control, which are today not really available for these technologies at all. The observation of structures on the nano-scale is difficult using visible light. Optical metrology, using radiation with high-energy photons, is a destructive metrology, inducing defects in the investigated object. Also, considering the ultra-fast velocity of chemical or molecular reactions in the picosecond and femtosecond regime, for instance the biochemical synthesis of chromosomes are not detectable using current diagnosis techniques and need new methods in addition. New powerful tools for the observation in the temporal and spatial domain are needed.

17) http://www.scanlab.de
18) For example see varioSCAN http://www.scanlab.de/index.php?id=17917
19) http://www.ilt.fraunhofer.de/eng/101006.html

2.4.1
Temporal Domain

The time-scales opened by ultra-fast laser radiation as a measuring tool are beyond the resolution of conventional detection techniques. The processes involved in CPUs (micro electronics), microelectromechanical systems MEMS (micro- and nanotechnology) are becoming ultra-fast. Today fiber network clock times are approximately some 10 GHz. This frequency is comparable to a time scale of several tens of picoseconds. Another example, the ultra-fast manipulation of matter needs deeper process understanding in order to fulfill ultra-precise manipulations, like cell-wall drilling for nano-invasive genomics, or DNA cutting. The involved processes induced by ultra-fast laser radiation are themselves of ultra-fast nature. Especially the interaction of matter with laser radiation:
- absorption of optical energy by the electrons lasts at least about 1 fs,
- interaction of the excited electrons with the electron system evolves within some 100 fs and with the phonon-system within some ps, and
- thermalization of the phonon-system by atomic collisions developed on a nanosecond time-scale.

As a consequence of ultra-fast process involved in ultra-fast engineering, an ultra-fast observation is needed.

Ultra-fast processes need ultra-fast and new diagnostics, because the conventional ones have insufficient resolutions regarding time-scales. Conventional optical diagnostics like photo-detectors, for example, a photo multiplier, photo diodes, and CCD, have minimal temporal resolutions in the time-scale of nanoseconds. Pump and probe techniques have been developed to detect processes on an ultra-fast time scale. This technique is the challenge for micro- and nanotechnology engineering.

2.4.2
Spatial Domain

The reachable resulting features induced by ultra-fast laser radiation can be beyond the optical resolution using conventional techniques like optical microscopy. Due to non-linear processes the effective laser focus can be reduced to well below the resolution of optical microscopy. Using femtosecond laser radiation enables routine structuring of biological material[20], cutting of neurons [80], generation of nanostructures like nanojets by melting [81], and selective polymerization of the resin[21] with a precision of about 50 nm (Figure 2.13). Additionally the processes induced by ultra-fast laser radiation run on a time scale where non-thermal equilibrium of matter is probable: melt dynamics in metals induced by pulsed femtosecond laser radiation is different induced by "cw" radiation.

20) http://darwin.bth.rwth-aachen.de/opus3/
volltexte/2009/2615/pdf/Wagner_Ralph.pdf
21) http://reichling.physik.uos.de/NanoForum/
pressillus.htm

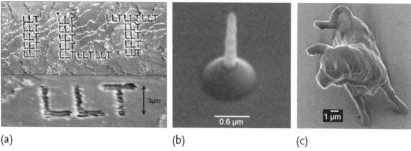

(a)   (b)   (c)

Fig. 2.13 Subwavelength structures on human hair[20] (a)
and nanojet fabricated in a 60 nm thick gold film [81] (b), and
three-dimensional nanoscopic objects generated by selective
polymerisation[21] (c) with femtosecond laser radiation.

In order to avoid, for example, harmful effects for surgery, the interaction of
ultra-fast laser radiation with matter has to be investigated and, in a second step,
controlled.

Even in fields where the laser radiation is not adopted as a machining tool, control by ultra-fast diagnostics is becoming increasingly necessary.

Conventional diagnosis for mechanical and electrical engineering applications
are insufficient for the exploding micro- and nanotechnologies. When structures
below the micrometer scale are generated, it demands consequently elaborate observation techniques like scanning electron microscopy (SEM), scanning tunnel
microscopy (STM), scanning near-field microscopy (SNOM) and atomic force microscopy (AFM).

2.5
Domains of Optical Pump and Probe Techniques for Process Diagnosis

Optical pump and probe metrology is described by a class of methods for the investigations of ultra-fast processes. Physical and chemical changes of matter, like
heating, melting, evaporation, oxidation, ionization, and so on, can be investigated
on a small time-scale using ultra-fast methods to probe the temporal evolution. The
fundamental principle is to excite a special characteristic of the involved process,
to make its time-dependence detectable by the investigation. The detectability of
a process is achieved by changing electronic properties like ionization state, oxidation number, phase of matter, optical phase, electron density, and electron kinetics
(for example, velocity). Three domains for pump and probe metrology can be distinguished.

- **Temporal domain:** Optical pump and probe technology is used for time-
  resolved detection of processes. The temporal resolution is given by the
  pulse duration of the pump and the probe laser radiation. This key property enables femtosecond, and even attosecond temporal resolution by us-

ing ultra-fast laser radiation. The temporal resolution of pump and probe diagnostics is larger than:

- the reaction times of chemical reactions between atoms,
- the isomerization times in molecules, and
- the inter-atomic time-scales for electron "movements". For example, the generation of Auger-electrons have been temporally resolved by using attosecond laser radiation [23].

Pump and probe techniques are used to process imaging or to get spatially resolved information of process parameters. Expansion of plumes and plasmas can be fixed in space and time with femtosecond resolution. The spatial resolution is defined by the focus size, which is controlled by the optical system used and the wavelength of the laser probe radiation (see Section 2.2). When multi-photon processes of the probe radiation are involved, the spatial resolution scales inversely proportional to the square root of the multi-photon index N

$$\Delta x \propto \frac{\lambda}{\sqrt{N}} . \tag{2.48}$$

- **Spectral domain:** Pump and probe techniques are used to get spectrally high-resolved and high-contrasted information on processes like chemical reactions. For one purpose the investigation system is pushed into a defined energetic state and then excited by the experimental system. This method, called excited state spectroscopy, is used often in chemistry. By using non-linear processes the probe radiation can be spectrally broadened to an ultra-fast white-light continuum. With this probe radiation the spectral information of a process over a large spectrum from UV to IR, can be detected instantaneously by spectrometry. Femtosecond laser radiation can be transformed by non-linear interaction with matter into high energy EUV-, X-radiation or into the radio regime, as THz radiation. The high-energy radiation of X-rays enables atomic and sub-atomic time resolved spectroscopy, whereas the T-radiation is applicable for material testing and security.

Depending on the information to be extracted from the experiment, only one of these domains may be of interest.

3
Fundamentals of Laser Interaction

Ultra-fast laser radiation interacting with matter exhibits three optical excitation principal processes:

- polarization of matter,
- ionization of matter, and
- excitation of free particles.

The first process describes the processes in linear and, where in general the state of matter is not changed (Section 3.1). High-power physics, being the second described process, is adopted when matter is transferred into the vaporized or plasma state (Section 3.2). The effects involved during the interaction of ultra-fast laser radiation with a plasma are the last presented process (Section 3.3).

3.1
Linear to Non-linear Optics

For dielectrics, laser radiation interacts predominantly with electrons at intensities $I < 10^{16}\,\mathrm{W\,cm^{-2}}$ below the interatomic electric field. Depending on the harmonic or non-harmonic oscillations of the electrons in the radiation field, linear or non-linear optics is adopted. Ultra-fast laser radiation is described by linear optics[22] for intensities $\ll 10^{10}\,\mathrm{W/cm^2}$ (Section 3.1.1). Whereas, for optical materials irradiated by ultra-fast laser radiation with intensities $\approx 10^{10}\,\mathrm{W/cm^2}$ exhibit a non-linear response to the irradiation, with the consequence that properties of the laser radiation, like pulse duration or spectral bandwidth, change (Section 3.1.2).

3.1.1
Linear Optics: Group Velocity Dispersion and Chirp

Induced by the electromagnetic field, the electrons of the valence band oscillate and induce dipoles with a total polarization **P**. In the case of intensities of radia-

[22] The equations adopted are written in the cgs
system common in the ultra-fast community

Ultra-fast Material Metrology. Alexander Horn
Copyright © 2009 WILEY-VCH Verlag GmbH & Co. KGaA, Weinheim
ISBN: 978-3-527-40887-0

tion where the displacements of the electrons are harmonic, the amplitude of the electromagnetic field is harmonic and a linear polarization $\mathbf{P}^L$ is induced in the dielectric. Reflection and refraction are examples of such *linear processes*. If the valence electrons oscillate by the electromagnetic field non-harmonically, a nonlinear polarization $\mathbf{P}^{NL}$ is induced. Self-focusing and self-phase modulation are examples of such *non-linear processes*. The polarization $\mathbf{P}$ can be split in a linear and in a non-linear term

$$\mathbf{P} = \mathbf{P}^L + \mathbf{P}^{NL} \,. \tag{3.1}$$

With the linear polarization $\mathbf{P}^L$ one describes the linear optics; therefore the non-linear polarization $\mathbf{P}^{NL}$ is neglected in the following considerations, also because objectives and lenses are designed to operate in the linear regime.

The propagation of electromagnetic radiation in a dielectric can be described by the Maxwell wave equation

$$\left(\nabla^2 - \frac{1}{c^2}\frac{\partial^2}{\partial t^2}\right)\mathbf{E}(\mathbf{r}, t) = \mu_0 \frac{\partial^2}{\partial t^2}\mathbf{P}(\mathbf{r}, t) \,. \tag{3.2}$$

For a linear plane wave $\mathbf{E} = E(z, t)\mathbf{e_x}$ propagating in the z-direction, Eq. (3.2) is thus reduced to

$$\left(\frac{\partial^2}{\partial z^2} - \frac{1}{c^2}\frac{\partial^2}{\partial t^2}\right)E(z, t) = \mu_0 \frac{\partial^2}{\partial t^2}P^L(z, t) \,. \tag{3.3}$$

The linear polarization written in the frequency space is given by

$$\tilde{P}^L(\omega, z) = \varepsilon\chi(\omega)\tilde{E}(z, \omega) \,, \tag{3.4}$$

described by the dielectric susceptibility tensor χ. By Fourier transformation, Eq. (3.2), the Maxwell wave equation becomes

$$\left[\frac{\partial^2}{\partial z^2} + \frac{\omega^2}{c^2}\varepsilon(\omega)\right]\tilde{E}(z, \omega) = 0 \,, \tag{3.5}$$

with the dielectric constant $\varepsilon(\omega) = [1 + \chi(\omega)]$. A plane wave is described in the frequency space by $\tilde{E}(\omega, z) = \tilde{E}(\omega, 0)\exp(-ik(\omega)z)$ with the wave number k following the dispersion relation

$$k^2(\omega) = \frac{\omega^2}{c^2}\varepsilon(\omega) = \frac{\omega^2}{c^2}n^2(\omega) \,. \tag{3.6}$$

To describe the dispersion of ultra-fast laser radiation with pulse durations in the femtosecond regime, the wave number

$$k(\omega) = k_l + \frac{dk}{d\omega}\bigg|_{\omega_l}(\omega - \omega_l) + \frac{1}{2}\frac{d^2k}{d\omega^2}\bigg|_{\omega_l}(\omega - \omega_l)^2$$

$$+ \frac{1}{4}\frac{d^3k}{d\omega^3}\bigg|_{\omega_l}(\omega - \omega_l)^3 \cdots = k_l + \delta k + k''' + O \quad (3.7)$$

is developed around a carrier frequency ω_l and described with the plane wave $\tilde{E}(\omega, z) = \tilde{E}(\omega, 0)e^{-ik_l z}e^{-i\delta kz}$. By introducing retarded spatial and temporal coordinates $\xi = z$ and $\eta = t - \frac{z}{v_g}$ with the group velocity of the pulse $v_g = (dk/d\omega|_{\omega_l})^{-1}$ for the Eqs. (3.5) and (3.7) a simplified wave equation

$$\frac{\partial}{\partial \xi}\tilde{\mathcal{E}}(\eta, \xi) - \frac{i}{2}k_l''\frac{\partial^2}{\partial \eta^2}\tilde{\mathcal{E}}(\eta, \xi) = 0 \tag{3.8}$$

can be derived. This equation describes the dispersion of laser radiation in matter in the impulse space [72]. The group velocity dispersion (GVD)

$$k_l'' = \left.\frac{\partial^2 k}{\partial \omega^2}\right|_{\omega_l} = \frac{2\pi c}{\omega^2 v_g^2}\frac{dv_g}{d\lambda} \tag{3.9}$$

describes the change of the group velocity v_g with the wavelength and therefore, as described in the following, the increase of the pulse duration of the laser radiation. Equation (3.8) in the frequency space is solved by $\tilde{E}(\omega, z) = \tilde{E}(\omega, 0)e^{-\frac{i}{2}k''\omega^2 z}$ and by inverse Fourier transformation follows the time-dependent complex electric field

$$\tilde{\mathcal{E}}(t, z) = \mathcal{F}^{-1}\left[\tilde{E}(\omega, z)\right] . \tag{3.10}$$

An electric field with a Gaussian frequency distribution $\tilde{\mathcal{E}}$ has a Gaussian-like temporal distribution in the impulse space

$$\tilde{\mathcal{E}}(t, z) = Ae^{-\left(1+i\frac{2k_l'' z}{t_p}\right)\left(\frac{t}{t_p(z)}\right)^2} . \tag{3.11}$$

The pulse duration

$$t_p(z) = t_{p0}\sqrt{1 + \left(\frac{2z\,|k_l''|}{t_p^2}\right)^2} = t_{p0}\sqrt{1 + \left(\frac{z}{L_D}\right)^2} \tag{3.12}$$

is increased by the group velocity dispersion k_l'' described by the characteristic dispersive length $L_D = \frac{t_{p0}^2}{2|k_l''|}$. The laser radiation is "chirped". The different frequency components of the laser pulse are temporally displaced from each other, whereas the temporal sequence of the frequency components remains linear.

The chirp $b = \frac{d\varphi}{dt}$ is defined as the derivation of the phase φ with the time. For a Gaussian temporal distribution the chirp can be written as

$$b = \frac{d\varphi}{dt} = -\frac{2a}{t_p^2} , \tag{3.13}$$

with a being the chirp parameter [72]. If the laser radiation is chirped the equation for the electromagnetic field is given by

$$\tilde{\mathcal{E}} = \mathcal{E}_0 e^{-(1+ia)\left(\frac{t}{t_p}\right)^2} . \tag{3.14}$$

The pulse duration-bandwidth product of the laser radiation is given by

$$\Delta\omega_p t_p = \sqrt{8\ln 2(1 + a^2)}. \tag{3.15}$$

The pulse duration is increased for a chirp parameter $a > 1$.

3.1.2
Non-linear Processes

3.1.2.1   **Non-linear Polarization**

Non-linear optics is the field of optics that describes the behavior of light in media in which the polarization **P** responds non-linearly to the electric field **E** of the radiation. This non-linearity is observable at large radiation intensities $> 10^{10}\,\mathrm{W\,cm^{-2}}$. For large intensities the non-linear response of the polarization of a dielectric can not be neglected anymore. To describe non-linear processes like self-focusing, the polarization

$$\mathbf{P} = \chi^{(1)}\varepsilon_0\mathcal{E} + \chi^{(2)}\mathcal{E}^2 + \chi^{(3)}\mathcal{E}^3 + \cdots \tag{3.16}$$

is Taylor expanded. $\chi^{(1)}$ is the linear susceptibility, and $\chi^{(i)}$ $(i > 1)$, are terms for the non-linear susceptibility of ith order. The non-linear susceptibilities are much smaller than the linear one and only become significant at large field strengths.

The second-order susceptibility $\chi^{(2)}\mathcal{E}^2$ describes the generation of the second harmonic (SHG) and the parametric interactions of laser radiation with matter, like the optical parametric amplification (OPA). The second-order susceptibility vanishes for centro-symmetric crystals. The SHG process involves two photons of the same frequency ω_1 interacting in the dielectric, generating one photon with the frequency $\omega_2 = 2\omega_1$. *Frequency mixing* describes the parametric interaction of three photons generating photons of the frequencies $2\omega_1$, $2\omega_2$, the sum-frequency $\omega_1 + \omega_2$ and the difference-frequency $\omega_1 - \omega_2$ [82].

By applying a voltage at a non-linear dielectric, the electro-optic effect of second-order, called *Pockels effect*, is induced, rotating the polarization of the incident radiation proportional to the applied voltage. In this way the refractive index n is changed

$$\Delta n_m \simeq -\frac{n^3}{2}\sum_p \frac{\chi^{(2)}_{pm}}{n^4}E_p \tag{3.17}$$

with $m = x, y, z$ being the orientation of the crystal axis. The Pockels effect is applied to modulate and to optically switch laser radiation. The optical rectification by a non-linear crystal is the inverse Pockels effect. In the case of interaction of laser radiation with this crystal, an electrical voltage is induced generating THz radiation by optical rectification.

The non-linear polarization $\mathbf{P}^{\mathrm{NL}}$ can be described as a function of the electric field in the third-order. The third-order susceptibility $\chi^{(3)}$ describes the generation of the third harmonic (THG), the Kerr effect, the self-focusing and self-phase modulation.

Birefringence is induced on an optical isotropic dielectric by applying an electric field represented by the electromagnetic field of intense laser radiation or an external applied electric field. The *Kerr effect* is achieved by the orientation of dipoles in the dielectric and inducing a refractive index change

$$\Delta n = n(E) - n_0 = \frac{1}{2}n_2\,|E|^2\,, \tag{3.18}$$

where n_2 is the second-order non-linear refractive index. Ultra-fast laser radiation passing through a dielectric, like lenses and complex objectives obtains

1. an increased pulse duration by dispersion (Chirp) and
2. an increased spectral bandwidth by non-linear processes like the Kerr effect.

Pulsed laser radiation with a Gaussian spatial intensity distribution interacting with a dielectric induces a refractive index change with a Gaussian spatial distribution, called gradient index. The laser radiation can be refracted and focused by this gradient index lens depending on pulse durations larger than the process time for this effect. The process is called *self-focusing*.

In the case of time-dependent electromagnetic fields, for example, pulsed laser radiation, the Kerr effect is also time-dependent. The spectral bandwidth is increased by *self-phase modulation*. For laser radiation with a Gaussian temporal distribution

$$I(t) = \exp\left[-(t/t_0)^2\right] ,$$ (3.19)

and a spatial distribution given by a plane wave

$$E(t, x) = E_0 \exp[\omega_0 t - kx] \quad \text{with the wave number} \quad k = n(t)\omega_0/c ,$$ (3.20)

results in a frequency ω of the radiation given by

$$\omega(t) = \frac{\partial \varphi(t)}{\partial t} = \omega_0 - \frac{\omega_0}{c} \frac{\partial n(t)}{\partial t} x .$$ (3.21)

The change in frequency $\delta \omega = -\frac{x \omega_0 n_2}{2c} \frac{\partial I(t)}{\partial t}$ describes new frequencies. The bandwidth of the laser radiation is increased [82].

3.1.2.2 Self-focusing
Self-focusing can be described in a macroscopic context of the polarization $\mathbf{P}$, Eq. (3.2), and by separation of the polarization in a linear and a non-linear term, Eq. (3.1). By introducing a linear and a non-linear polarization the wave Eq. (3.2) will be simplified. The linear term

$$\mathbf{P}^{\mathrm{L}}(x, t) = \int_{-\infty}^{\infty} \hat{\chi}(t - t') \cdot \mathbf{E}(x, t')dt'$$ (3.22)

describes the retarded polarization [83]. This description is valid for homogeneous optical non-active media. The causality determines the retarded polarization, Eq. (3.22), non-negative values for the susceptibility $\hat{\chi}$. The Fourier transformed polarization can be written as

$$\mathbf{P}^{\mathrm{L}}(\mathbf{x}, \omega) = \chi(\omega) \cdot \mathbf{E}(\mathbf{x}, \omega) .$$ (3.23)

By replacing $\tau = t - t'$ in Eq. (3.22) and writing the electric field as a complex electric field

$$\mathbf{E}(\mathbf{x}, t - \tau) = \left[\exp i\left(i \frac{\partial}{\partial t}\right)\tau\right] \mathbf{E}(\mathbf{x}, t) ,$$ (3.24)

the complex linear polarization becomes $P^L(x, t) = \chi \left(i\frac{\partial}{\partial t} \right) \cdot E(x, t)$. In one dimension the linear complex polarization results in

$$\mathcal{P}^L(x, t) = \chi \left(\omega_0 + i\frac{\partial}{\partial t} \right) \cdot \mathcal{E}(x, t) \tag{3.25}$$

for nearly monochromatic light in SVE approximation (Slow-Varying-Envelope) with the complex electric field $\mathcal{E}(x, t) = \mathcal{E}_0 \exp(-i(\omega_0 t - k_0 z))$. Developing the linear polarization from the electric field $\mathcal{E}$ considering only terms up to the second-order in $\mathcal{E}$ and in higher order with respect to z and t by the SVE approximation, one obtains an approximative wave equation

$$2ik_0 \left(\frac{\partial}{\partial z} + \frac{1}{v_g}\frac{\partial}{\partial t} \right) \mathcal{E} + \nabla_T^2 \mathcal{E} = -\frac{4\pi}{c^2}\omega_0^2 \mathcal{P}^{NL} - \frac{8\pi i \omega_0}{c^2}\frac{\partial \mathcal{P}^{NL}}{\partial t} \tag{3.26}$$

with the group velocity defined by $v_g = \left(\frac{d}{d\omega}\frac{n\omega}{c} \right)^{-1}$.

Self-focusing can be described by P^{NL}, whereas the induced dipoles are oscillating at frequencies close to the frequency of the laser radiation ω_0. The non-linear third-order polarization is defined by

$$\mathcal{P}_j^{NL}(\omega_4) := D\chi_3^{jklm}(-\omega_4, \omega_1, \omega_2, \omega_3)\mathcal{E}_k(\omega_1)\mathcal{E}_l(\omega_2)\mathcal{E}_m(\omega_3) \tag{3.27}$$

and can be resolved for monochromatic light to

$$\mathcal{P}_j^{NL} = \eta |\mathcal{E}|^2 \mathcal{E}_j, \tag{3.28}$$

with the crystal directions $j = x, y$ and the susceptibility factor $\eta \propto \chi_3^{jklm}$. The refractive index change Δn results in

$$\Delta n = \frac{2\pi\eta}{n} |\mathcal{E}|^2 = \frac{16\pi^2}{n^2 c}\eta I, \tag{3.29}$$

using the intensity $I = nc |\mathcal{E}|^2 / 8\pi$ (cgs system) and the non-linear refractive index $n_2 = 4\pi\eta/n$ with the Eq. (3.18).

Without a time dependence in Eq. (3.26) and taking a scalar electric field ($\nabla(\nabla \cdot E) = 0$), the wave equation reduces to

$$2ik_0 \frac{\partial \mathcal{E}}{\partial z} + \nabla_T^2 \mathcal{E} = -\frac{4\pi\omega_0^2}{c^2} \mathcal{P}^{NL}. \tag{3.30}$$

Using a Hamilton–Jacobi equation the wave equation can be analytically solved after separation of the wave function in an amplitude and a phase function [83]. The numerical solution of Eq. (3.30) is described for $P > P_{c1}$, with the threshold for self-focusing P_{c1} and the position of the self-focus z_{fs} as a function of the peak power

$$\left(\frac{P}{P_{c1}} \right)^{0.5} = 0.852 + 0.365 \frac{z_R}{z_{sf}}. \tag{3.31}$$

Equation (3.31) describes the modified caustic of the laser radiation by self-focusing.

Pulsed laser radiation (pulse duration $t_p \leq 200\,\text{fs}$) with a spatial Gaussian intensity distribution propagating in vacuum is described for cw-laser radiation [72] by

$$\tilde{w}(x, y, z) = \frac{w_0}{\sqrt{1 + z^2/z_R^2}} e^{-i\Theta(z)} e^{-ik_l(x^2+y^2)/2\tilde{q}(z)}, \tag{3.32}$$

with $\Theta(z) = \arctan(z/z_R^2)$ being the phase and z_R the Rayleigh length, Eq. (2.10). w_0 represents the beam radius and λ_l the wavelength of the applied laser radiation. $\tilde{q}(z)$ is the complex beam parameter

$$\frac{1}{\tilde{q}(z)} = \frac{1}{R(z)} - \frac{i\lambda_l}{\pi w^2(z)} = \frac{1}{q(0) + z} \tag{3.33}$$

with $R(z) = z + z_R^2/z$ being the phase curvature and $w(z) = w_0\sqrt{1 + z^2/z_R^2}$ the beam radius at the position z. The spatial intensity distribution in the propagation direction (caustic) at intensities below the threshold for self-focusing P_{c1} is given by

$$I(z) = \frac{I_0}{\left(1 - \frac{z}{R}\right)^2 + \left(1 - \frac{P}{P_{c1}}\right)\left(\frac{z\lambda}{w_0^2\pi}\right)^2}, \tag{3.34}$$

where P represents the power of the laser radiation, $I_0 = I(z = 0)$, and R the curvature of the phase front at the surface of the dielectric. By determining the maximum of $I(z)$ (Eq. (3.34)) the position of the focus is calculated

$$\frac{\partial I(z)}{\partial z} \overset{!}{=} 0 \longrightarrow z_{I_{max}} = \frac{\pi^2 R w^4}{\pi^2 w^4 + R^2\lambda^2 - \frac{P}{P_{c1}}R^2\lambda^2}. \tag{3.35}$$

The calculations are valid for self-focusing of cw-laser radiation. Assuming that self-focusing is an instant process, the power can be replaced by the peak power. This assumption is justifiable for the non-resonant Kerr effect, occurring in crystals and glasses for pulse durations $t > 50\,\text{fs}$, because the non-resonant process for self-focusing evolves on a time scale of some femtoseconds [72]. This has the consequence that the focus moves in the direction of the laser during irradiation of dielectrics for pulse durations above 50 fs.

3.1.2.3 Spectral Broadening by Self-phase Modulation

Ultra-fast laser radiation with a Gaussian temporal distribution and a constant phase is given by

$$I(t) = I_0 \exp\left(-\frac{t^2}{t_p^2}\right) \tag{3.36}$$

where I_0 is the peak intensity, and t_p the pulse duration.

If the pulse is traveling in an optical medium with the linear refractive index n_0, and the non-linear refractive index n_2, by the optical Kerr effect an intensity-dependent refractive index change is induced

$$n(I) = n_0 + n_2 \cdot I(t). \tag{3.37}$$

The intensity at any one point in the medium rises and then falls as a function of time and induces a time-varying refractive index

$$\frac{dn(I)}{dt} = n_2 \frac{dI}{dt} = n_2 \cdot I_0 \cdot \frac{-2t}{t_p^2} \cdot \exp\left(\frac{-t^2}{t_p^2}\right),$$

(3.38)

and induces a shift in the instantaneous phase of the pulse:

$$\phi(t) = \omega_0 t - \frac{2\pi}{\lambda_0} \cdot n(I) L$$

(3.39)

where ω_0 and λ_0 are the carrier frequency and wavelength of the pulse, and L is the propagation distance through the optical medium. The phase shift results in a frequency shift of the pulse. The instantaneous frequency $\omega(t)$ is given by

$$\omega(t) = \frac{d\phi(t)}{dt} = \omega_0 - \frac{2\pi L}{\lambda_0} \frac{dn(I)}{dt},$$

(3.40)

and from the Eq. (3.38) follows

$$\omega(t) = \omega_0 + \frac{4\pi L n_2 I_0}{\lambda_0 t_p^2} \cdot t \cdot \exp\left(\frac{-t^2}{t_p^2}\right).$$

(3.41)

The leading edge of the pulse shifts to lower frequencies ("redder" wavelengths), the trailing edge to higher frequencies ("bluer") and the very peak of the pulse is not shifted. For the center portion of the pulse (between $t = \pm t/2$), there is an approximate linear frequency shift (chirp) given by:

$$\omega(t) = \omega_0 + \left.\frac{d\omega}{dt}\right|_0 \cdot t = \frac{4\pi L n_2 I_0}{\lambda_0 t_p^2}.$$

(3.42)

Additional frequencies have been generated through SPM and broaden the frequency spectrum of the pulse symmetrically. In the time domain, the envelope of the pulse is not changed.

The self-phase modulation is very efficient for ultra-fast laser radiation, because this process is inversely proportional to the pulse duration t_p. The resulting spectral distribution is symmetric. Considering larger terms of the non-linear polarization $\mathbf{P}^{NL}$, an asymmetric spectral distribution of the white light continuum can be calculated corresponding to the experimental results [84]. The spectral broadening is induced by self-phase modulation and also by four-wave mixing. The relative broadening of the white light continuum can be calculated for a temporal $sech^2$ intensity distribution $I(\mathbf{r}, \tau) \propto sech^2(\tau/t_p)$ to

$$\frac{\Delta\omega_{\pm}^{SPM}}{\omega_0} = \frac{1}{2}\left(\sqrt{Q^2 + 4} \pm |Q|\right) - 1,$$

(3.43)

with $Q = 2n_2 IL/ct_p$. For $Q \ll 1$ Eq. (3.43) merges into Eq. (3.45).

The reason for the experimentally observed asymmetry of the spectral distribution of a white light continuum is explained by the temporal and spatial change in the pulse shape. The pulse shape is spatially modified by self-focusing and temporally by self-phase modulation. The fact that the phase change has first a positive value and then a negative one results in different velocities of individual parts of the pulse. The temporal distribution change of the pulse is called *self-steepening*.

3.1.2.4 Catastrophic Self-focusing

A white light continuum is generated by focusing ultra-fast laser radiation into a dielectric. Different processes are taking place during this process:

- self-focusing,
- self-steepening, and
- self-phase modulation.

Filaments are formed in a dielectric by self-focusing and self-defocusing of the laser radiation. For intensities beyond a material specific threshold, free electrons are generated by multi-photon absorption or avalanche ionization. These electrons interact with the laser radiation like a dispersing lens, because the refractive index of an electron gas is $n_e < 1$, and the spatial electron distribution $\varrho(r)$ corresponds to the Gaussian spatial intensity distribution of the laser radiation. The change in refractive index

$$\Delta n = -\frac{2\pi e^2 N_e}{n_0 m_e (\omega_0^2 + v_e^2)}, \tag{3.44}$$

generated by free electrons, can be described by the Drude model [85], where N_e represents the electron density, v_e the electron collision frequency and ω_0 the frequency of the radiation. The refractive index change induced by the Kerr effect Δn_{Kerr} is abrogated for electron densities $N_e \approx 10^{17}$–10^{18} cm^{-3} by the refractive index change induced by the free electrons.

The intensity of the laser radiation increases by self-focusing and self-phase modulation becomes dominant. With self-phase modulation additional phase terms in the polarization are induced by the electromagnetic field of the laser radiation

$$\varphi_{NL}(\tau) = \int_0^L n_2 I(z, \tau) \frac{\omega_0}{c} dz. \tag{3.45}$$

L is the thickness of the dielectric. The additional frequency terms of the spectrum, which are generated by the additional phase terms, contain frequencies from the minimal value of Stokes frequency

$$\Delta\omega_-^{SPM} = \left(-\frac{d\varphi_{NL}}{d\tau}\right)_{min} \tag{3.46}$$

to the maximal value of the anti-Stokes frequency

$$\Delta\omega_+^{SPM} = \left(-\frac{d\varphi_{NL}}{d\tau}\right)_{max}. \tag{3.47}$$

Temporal modulation of the intensity distribution of the laser radiation induces a change in the focal position, Eq. (3.35). During irradiation with a spatially and temporally localized pulse propagating through the dielectric, the focal position is increasingly displaced with increasing intensity of the pulse towards the source. With decreasing intensity the displacement has a negative sign. This process of *moving focus* takes place during the pulse duration of the laser radiation τ_p for $\tau_p \gg \tau$, with τ being the process time.

In cases of de-focusing and self-focusing in a dielectric resulting in free electrons, the self-phase modulation can be solved using the non-linear Schrödinger equation, considering the multi-photon ionization and the moving focus [86]. Solutions of the Schrödinger equation postulate a frequency-dependent angle distribution of the white light continuum [87].

3.2
High-Power Photon–Matter Interaction

In general, the interaction of high-power pulsed laser radiation with solid materials involves complex phenomena. Optical, thermal, mechanical or hydrodynamic processes can be present contemporaneously during drilling. During the first instance of the interaction, the incident optical energy is distributed among these phenomena. For the investigations on metals, using ultra-fast laser radiation in the near-infrared wavelength range, linear and two-photon absorption is assumed to be the dominant absorption mechanism [88–90]. Part of the initial energy is absorbed leading to the heating of the bulk. Then the melting and the vaporization of the irradiated sample can be observed.

Intense ultra-fast laser radiation interacting with matter is partly absorbed, scattered or reflected (Section 3.2.1). The absorbed laser radiation induces heating of the matter. The time-scales for absorption into the electron system and subsequent heating of matter by the interaction of the electron system with the phonon system are different and have to be considered when ultra-fast laser radiation interacts with matter (Section 3.3.2). The temperature of the material can rise above melting and evaporation temperatures. The time-scales for melting and vaporization are given for metals and dielectrics (Section 3.2.3). A short description of matter ionization is given in Section 3.2.6.

The processes involved during laser radiation interaction with metals and with dielectrics are described in the following separately, because of the different physical properties of metals and dielectrics.

3.2.1
Absorption of Laser Radiation in Matter

In contrast to heating of matter by long-pulsed laser radiation due to a thermalized electron and phonon system, using femtosecond laser radiation, absorption and heating are non-equilibrium processes. The radiation is absorbed by bound and free electrons of metals. For intensities $I > I_{thr}$, where I_{thr} represents the threshold intensity for ionization, absorption is accompanied by ionization of the material.

Femtosecond laser radiation heats a solid target faster than the time scale for hy-drodynamic expansion, so that the density of the solid remains nearly unchanged. Laser irradiation with pulse durations below $t_p \approx 100\,fs$ results in heated skin of about 20 nm thickness. In this time, a generated plasma expands by only 1 nm [91].

For a one-dimensional approach the absorption coefficient A at the surface is given by

$$A = \frac{\int_0^{2t_p} dt \int_0^{\infty} Q(z, t)dz}{\int_0^{2t_p} I(t)dt}, \tag{3.48}$$

with the temporal intensity distribution of the laser radiation approximated by $I = I_0 \sin^2 \left(\frac{\pi t}{2t_p} \right)$ for $0 \leq t \leq 2t_p$ and $I = 0$ elsewhere and the pulse duration t_p (FWHM). The Joule heating Q is given by

$$Q(z, t) = \frac{1}{2} \operatorname{Re}\{\sigma_E(z, t)\}|E|^2, \tag{3.49}$$

with the electrical conductivity σ_E containing the contribution for inter-band (band-band transition) and intra-band absorption (inverse Bremsstrahlung)

$$\sigma_E = \sigma_{bb} + \sigma_D. \tag{3.50}$$

The inverse Bremsstrahlung written as the Drude contribution is

$$\sigma_D = \frac{n_e e^2}{m_{\text{eff}}} \left(\frac{\nu + i\omega}{\nu^2 + \omega^2} \right). \tag{3.51}$$

The inter-band transitions considered here occur between the Bloch electron bands. In metals with (partially) occupied atomic d sub-shells, a different class of inter-band transitions is also possible; namely, the transitions between the oc-cupied d-states and the Fermi surface. Inter-band transitions were considered in [92, 93] and even can contribute significantly to the absorption of ultra-fast laser radiation [94], but will not be treated here anymore.

The incident laser radiation delivered to the sample is either reflected, transmit-ted or absorbed. Absorptivity, A, refers to the fraction of the incoming radiation that is absorbed by the material.

The absorptivity of a dielectric material depends on the intensity of the laser ir-radiation and can vary with time during the duration of the laser pulse. A linear absorption of radiation in dielectrics is improbable, because the band gap of a di-electric is in general much larger than the adopted photon energy. The optical en-ergy can be deposited into a dielectric by absorption at defects or by multi-photon absorption in the dielectric. When glass is exposed to high intensity ultra-short pulses, its reflectivity increases with time as the plasma density increases. Once the critical surface plasma density is formed, any further incident laser energy is reflected from the surface due to an induced skin effect. At a laser fluence of $20\,J/cm^2$, the reflectance has been estimated to be about 60% (for $\lambda = 1064$ nm and $t_p = 350\,fs$) [95].

3.2.2
Energy Transfer from Electrons to Matter

In a metal irradiated by fs laser radiation, the temperature T_e of the electrons in the conduction band and the temperature T_i of the ion lattice (in other words, of the phonons) can differ by orders of magnitude and is due to a relatively low rate of energy exchange on this time scale between the electrons (which absorb the laser radiation) and the lattice. The rate of energy exchange within the electron system, on the other hand, is on the order of $10^{14}\,\mathrm{s}^{-1}$ or higher for $T_e \geq 1\,\mathrm{eV}$. Thus, the electrons may be considered to be in a local thermodynamic equilibrium which may be characterized by a certain temperature.

Irradiation of metals with small fs-laser intensities results in $T_e < 1\,\mathrm{eV}$, and the absorption coefficient remains essentially unchanged. The effects of the delayed thermalization of conductivity electrons may be detected by transmission spectroscopy near the absorption edge [93], but in absolute terms, the effect on the absorption coefficient A of the electron distribution deviation from equilibrium is quite small. At the temperatures considered the phonons may be characterized by a temperature T_i.

Following absorption the energy is deposited in the electron system. One description of the energy transfer from the electron to the phonon system has been proposed by Anisimov by the two-temperature model (TTM) [96]. Alternatively one can calculate by molecular dynamics (MD) the kinetics of each atom with the drawback of very limited interaction volume counting some 10^6 atoms [97–99].

Thus, in order to model the dependence of the absorption coefficient on the laser intensity, the following set of equations [96], called the two-temperature model (TTM), may be used

$$C_e(T_e)\frac{\partial T_e}{\partial t} = \frac{\partial}{\partial z}\left(\kappa(T_e)\frac{\partial T_e}{\partial z}\right) - U(T_e, T_i) + Q(z, t) \tag{3.52}$$

$$C_i(T_i)\frac{\partial T_i}{\partial t} = U(T_e, T_i) \tag{3.53}$$

where C_e and C_i are the electron and ion heat capacities at constant volume, κ is the electron heat conductivity and U

$$U = \gamma(T_e - T_i) \tag{3.54}$$

the energy transfer rate from electrons to ions described by the electron–phonon coupling constant γ. To calculate κ and U as function of T_e and T_i a knowledge of the phonon spectrum, in other words the dispersion relation, is needed as well as the electron–phonon interaction matrix element [94].

Thermalization between the electron and phonon system takes place during the irradiation when the pulse duration is much larger than the lattice heating time

$$\tau_i = \frac{C_i}{\gamma} \tag{3.55}$$

with typical values $\tau_i \approx 0.01$–1 ns. Then a thermal equilibrium can be given for the electron and phonon system with one temperature $T = T_i = T_e$ and the TTM simplifies to

$$C_i \frac{\partial T}{\partial t} = \frac{\partial k_e}{\partial x} \frac{\partial T}{\partial x} + I(t) A\alpha e^{-\alpha x}, \tag{3.56}$$

using as source term $Q = I(t)A\alpha e^{-\alpha x}$ with the absorptivity A and the absorption coefficient α [100].

Using femtosecond laser radiation, the energy transfer to the lattice during irradiation and the heat conduction can be neglected in a first approximation. As shown by [101–103] in this simplified case the ablation rate depends on the optical penetration depth only.

The absorbed energy is transferred for dielectrics from the high energy electrons to the lattice through electron–phonon scattering within the region of energy deposition. This happens within the first 10–20 ps [104]. Substantial heat diffusion begins only after a few tens of nanoseconds. In the case of femtosecond laser ablation of glass, a rather complicated and time consuming numerical solution is required to estimate the partition of the absorbed laser energy. Ben-Yakar *et al.* [104] proposed to use a measured *effective optical light penetration depth* to estimate the fraction of the incoming laser energy deposited in the glass as heat (thermal energy). The absorbed laser energy is deposited in a layer defined by the penetration of light. To quantify the penetration depth of radiation the Beer–Lambert law is adopted. The attenuation of the absorbed laser fluence as a function of depth is given by

$$F_a(z) = AF_0 \exp\left(-\frac{z}{\alpha_{\text{eff}}^{-1}}\right), \tag{3.57}$$

where the surface absorptivity for glass is $A = 0.3$–0.4 and the effective optical penetration depth is $\alpha_{\text{eff}}^{-1} = 238$ nm, as measured [105]. Three distinct regions for the absorption depth can be defined (Figure 3.1):

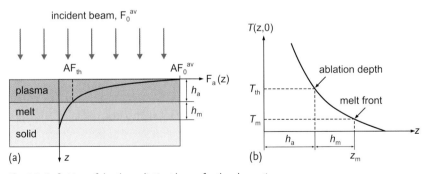

Fig. 3.1 Definition of the three distinct layers for the absorption depth (a), and temperature as function of depth (b) [104].

1. Ablation: the fluence of the absorbed radiation at the surface drops to the ablation threshold value (AF_{thr}) at the ablation depth. In this region, a high-pressure and high-temperature plasma is formed.
2. Melting: fluence drops below the ablation threshold, optical breakdown of glass cannot occur (electron number density is below the critical value) and the glass melts.
3. Heating: the fluence is too small and heating is insufficient to melt the glass.

3.2.3
Melting and Vaporization

Energetic free electrons overcoming the work function of the metal can escape from the target. Consequently, the electric field created by the charge separation between the escaping electrons and parent ions effectively could pull ions out of the target [106, 107]. Thus, the energetic fast ions [108] are likely emitted via a non-thermal process during the early stage of ablation in the time range up to some nanoseconds. It should be noted, however, that the net surface charge density due to the non-thermal process is two orders of magnitude smaller than in dielectric materials [109]. Gradual increase in absorption of the irradiated area was observed at small elapsed times $\tau < 3.0\,\text{ns}$ due to the emergence of a thin plasma layer in the proximity of the surface that attenuates the reflected probe laser beam used for diagnostics.

The plasma and melt dynamics can be described qualitatively well with some well-known theoretical issues [110] based on molecular dynamics assumptions. As shown in the experiment described below, however, unlike in the performed calculations, significant differences in the time-scales of dynamic processes can be observed due to focusing conditions and pulse energy applied differently. According to theoretical and experimental data on ablation dynamics of solids [111, 112], the expansion starts with a rarefaction wave that proceeds from the surface into the material. After reflection at the unperturbed substrate, the rarefaction wave travels back toward the sample surface. This results in a thin layer of high density, which moves away from the target in front of a low density region. The low density region is assumed to be a liquid–gas mixture caused by homogeneous nucleation within a few tens of picoseconds.

Up to pulse durations $t_p = 1\,\text{ps}$ only the electron system of matter is heated, whereas the lattice temperature increase is delayed up to 5 ps. The calculation using Eqs. (3.52) and (3.53) for heating and a QEOS model, as well an analytical approach for vaporization [113], demonstrate that after transfer of the energy to the lattice, vaporization starts taking away a major fraction of energy (Figure 3.2a). A considerable amount of energy remains in the lattice causing thermal load by heating and melting [108, 114].

Calculation for metals reveal that the expectation of cold machining by using ultra-fast laser radiation is not fulfilled completely [108]. The phonon system remains cold during irradiation. A much larger process time (than in the nanosecond regime) for solidification, formation of the maximal melt depth and evaporation

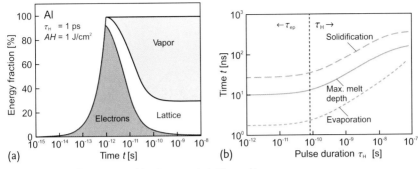

(a) (b)

Fig. 3.2 Energy fraction stored in electrons, in lattice and in vapor after absorption of laser radiation with 1 ps pulse duration (a), and characteristic times for evaporation, maximal melt depth and solidification (b) [108].

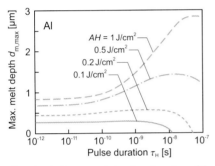

Fig. 3.3 Maximum melt depth as function of the pulse duration for different fluences [114].

results for irradiation of metals with picosecond laser radiation (Figure 3.2b). For pulse durations below the electron–phonon coupling time $\tau_{e-p} \approx 10^{-10}$ s the characteristic times for evaporation, melting and solidification saturate. A non-evanescent melt depth results, which decreases with decreasing pulse duration and fluence (Figure 3.3). A melt depth > 0.2 µm is detectable even for irradiation of metal with femtosecond laser radiation.

Ultra-fast laser ablation of dielectric materials such as glass involves a number of processes, including non-linear absorption, generation of plasmas, shock propagation, melt propagation and re-solidification. Each of these processes has a different time scale and can roughly be grouped into three different time domains (Figure 3.4):

- Part of the incident laser energy is absorbed in the picosecond regime by electrons through multi-photon and avalanche ionization and then transferred to the lattice on the time scale of few picoseconds (Figure 3.4a). As the electrons and phonon system thermally equilibrate, a high-pressure and high-temperature plasma is formed above the surface.
- In the nanosecond regime, the plasma expands primarily in the direction perpendicular to the target surface (Figure 3.4b).

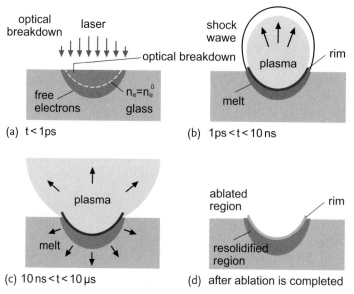

Fig. 3.4 Principle of the ablation process of dielectrics with femtosecond laser radiation for different time regimes: picosecond (a), nanosecond (b), microsecond regime (c), and ablated region (d) [104].

- In the microsecond regime, the plasma expands in both the lateral and perpendicular directions and removes the ablated material from the surface (Figure 3.4c).

3.2.4
Melt Dynamics

The thermal energy deposited in the bulk of a dielectric such as glass forms a transient shallow molten zone below the expanding plasma [105]. Ladieu *et al.* [115], for example, measured that about 8% of the incoming energy was thermalized and transmitted to the undamaged part of a quartz material when irradiated with a 100 fs laser pulse. During plasma expansion, the front of the molten material propagates into the bulk as a result of the heat diffusion. When the temperature of the melt decreases below the melting temperature the melt re-solidifies. The forces acting on the molten material drive the liquid from the center to the edges of the crater during the melt lifetime and create an elevated rim around the ablated crater as the melt re-solidifies.

Two main forces might affect the flow of a molten layer below the expanding plasma [104]:

1. Thermo-capillary forces (Marangoni flow), and
2. Forces exerted by the pressure of the plasma above the surface.

Thermo-capillary flow is induced on the surface following the Gaussian beam intensity profile of the laser radiation by the induced temperature gradient. The temperature gradient on the surface creates surface tension gradients that drive material from the hot center to the cold periphery. This response is expected in metals where the surface tension γ_s, decreases as the fluid becomes hotter ($d\gamma_s/dT < 0$). For glass $d\gamma_s/dT > 0$. Such a thermo-capillary flow in laser irradiated glass surfaces would actually drive fluid from the cold periphery to the hot center of the melt. In contrast, the experiments result in a flow of the melt toward the periphery, maybe by plasma recoil pressure [104]. In addition, the effect of thermo-capillary flow in glass is expected to be negligible because of its high viscosity, which leads to a flow time scale much larger than the typical melt time scale.

Pressure gradients inducing hydrodynamics exerted by the plasma onto the molten material may induce a gradient of ablation pressure on the molten surface and, as a consequence, a lateral melt flow to the periphery. The pressure gradients are particularly large at the plasma/air interface inducing a melt flow towards the periphery and a rise of a thin rim at the edges of the melted surface much like a splash of a liquid.

Melting may occur as a consequence of the interaction of femtosecond laser radiation with dielectrics has been demonstrated for fused silica and α-quartz [116]. The incubation effect in fused silica leads to high-temperature and high-pressure conditions which principally favor the phase transition (crystallization) of amorphous fused silica to crystalline quartz. In the case of the interaction with α-quartz, only amorphous re-solidified melt is observed. This could be due to the higher heat conductivity and thermal expansion coefficient of crystalline quartz compared to fused silica.

The hydrodynamics of a thin-film has been modeled with a two-dimensional model for the fluid motion and the imposed plasma pressure [104]. Two characteristic time scales have been derived:

- Marangoni flow

$$\tau_M \approx \frac{\mu L^2}{\gamma_T T_m \langle h_m \rangle} \tag{3.58}$$

- Pressure-driven flow

$$\tau_P \approx \frac{\mu L^2}{\langle p_{pl} \rangle \langle h_m \rangle^2} \tag{3.59}$$

where $\langle h_m \rangle$ is an average melt depth, L is a typical radial dimension and $\langle p_{pl} \rangle$ is an average plasma pressure, $\gamma_T = d\gamma_s/dT_S$ is a constant, and μ the viscosity of the melt. Calculations show that the characteristic time scale for Marangoni[23] flow is about three orders of magnitude larger than for pressure-driven flow, and $\tau_P \ll \tau_M$, even if the peak pressure is lowered more than a factor of ten. It is clear from this

---

23) The Gibbs–Marangoni effect is the mass transfer on, or in, a liquid layer due to surface tension differences.

estimation that the large plasma pressure above the free surface exerts more melt dynamics than the surface tension gradients [104].

3.2.5
Onset of Ablation

Considering heat capacity and thermal conductivity as constant, two geometrical regimes can be found where the temperatures of the phonons differ:

- In the first regime the optical penetration depth $\delta = 1/\alpha$ exceeds the thermal diffusion length l_{th}: $\delta > l_{th}$,
- In the second one, vice versa: $\delta < l_{th}$.

As a consequence, a threshold for ablation in depth x can be defined

$$F_a(x) \geq F_{th} \exp\left(\frac{x}{i}\right), \quad i = \delta, l \tag{3.60}$$

describing two distinct ablation rates [3, 117] (Figure 3.5). Because of strong non-equilibrium between the electron and phonon system, the thermal diffusion length l becomes smaller than the optical penetration depth for fluences $F \geq 0.5\,J/cm^2$. Irradiation of metals with ultra-fast laser radiation induces strong overheating of the electron system $T_e \gg T_i$ and the electron–phonon relaxation time becomes larger than the electron's relaxation time $\tau_{e-ph} > \tau_{e-e}$. The thermal conductivity can be approximated by $\kappa \sim T_e^{-1}$ and the thermal diffusivity by

$$D = \kappa/C_e \sim T_e^2. \tag{3.61}$$

With increasing electron temperature by increasing laser fluence, the thermal diffusivity and the electron thermal losses decrease.

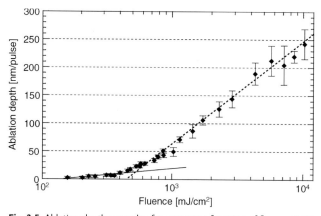

Fig. 3.5 Ablation depth per pulse for copper as function of fluence [117].

3.2.6
Ionization

Free electrons are generated after irradiation of matter above the threshold for photoionization and impact ionization. For large pulse durations electrons are generated by impact ionization and for small pulse durations by avalanche ionization involving photoionization [118]. In avalanche ionization, a free electron is heated in the large electric field until it has sufficient energy to ionize a second one. These two electrons repeat the process, and the electron number N grows exponentially following

$$\frac{dN_e}{dt} = \gamma_e(E)N_e \tag{3.62}$$

with γ_e the avalanche ionization coefficient, N_e the conduction band electron density, and the final electron density

$$N = N_e \exp(\gamma_e t). \tag{3.63}$$

Electrons absorbing optical energy generate ulterior free electrons by impact with atoms. Avalanche ionization results when free electrons continue to absorb optical energy from the electromagnetic field [119–121].

The coupling of strong electro-magnetic fields (like femtosecond laser radiation) with electrons can be described by the Boltzmann equation, which does not use phenomenological equations [122, 123]. One example is the simplified Fokker–Planck equation describing the electron density distribution $N_e(\varepsilon, t)$ of excited dielectrics

$$\frac{\partial N_e(\varepsilon, t)}{\partial t} + \frac{\partial}{\partial t}\left(R_J(\varepsilon, t)N_e(\varepsilon, t) - \gamma(\varepsilon)E_p N_e(\varepsilon, t) - D(\varepsilon, t)\frac{\partial N_e(\varepsilon, t)}{\partial \varepsilon}\right)$$
$$= S(\varepsilon, t), \quad (3.64)$$

with ε the kinetic energy of the electrons [124, 125]. The electron number with kinetic energy in the range ε to $\varepsilon + d\varepsilon$ is given by $N_e(\varepsilon, t)d\varepsilon$. R_J is the Joule resistance which describes the energy transfer by electron–phonon impact. $\gamma E_p N_e$ describes the energy transfer between electrons and phonons and $D\partial N_e/\partial \varepsilon$ is the energy diffusion constant for electrons. γ is the Keldysh parameter for photoionization and S describes sources and drains of electrons. The Eq. (3.64) describes the formation of electrons by avalanche and multi-photon ionization (in the case of dielectrics). The formation of free electrons by femtosecond laser radiation, the induced mechanical stress and the formation of shock waves in the bulk can be calculated by the Fokker–Planck Eq. (3.64), which can be simplified to

$$\frac{dN_e}{dt} = \alpha_a I(t)N_e(t) + \sigma_k I^k, \tag{3.65}$$

with $I(t)$ being the temporal-dependent intensity, α_a the avalanche ionization coefficient and σ_k the k-photon's absorption cross-section.

The photoionization can be described by the Keldysh equation

$$w_{PI}(E) = \frac{2\omega}{9\pi} \left(\frac{\omega m}{\sqrt{\gamma_1}\hbar} \right)^{3/2} Q(\gamma, x) \exp \left[-\pi \langle x + 1 \rangle \frac{\mathcal{K}(\gamma_1) - \mathcal{E}(\gamma_1)}{\mathcal{E}(\gamma_2)} \right] . \qquad (3.66)$$

The Keldysh parameter is defined as $\gamma = \frac{\omega \sqrt{m\Delta}}{eE}$ with the frequency of the radiation ω, the reduced mass of the electrons and holes m, the inter-band gap Δ, the electron charge e and the electric field strength E. $Q(\gamma, x)$ is a complex functional [126]. $\mathcal{E}$ and $\mathcal{K}$ represent elliptical integrals of first- and second-order and the coefficient γ_i $(i = 1, 2)$ is given by $\gamma_1 = \frac{\gamma^2}{1+\gamma^2}$ and $\gamma_2 = 1 - \gamma_1$. For small frequencies or intensities with $\gamma \ll 1$ the Keldysh equation merges into tunnel ionization. $\gamma \gg 1$ describes multi-photon ionization.

3.3
Matching of Plasma Dynamics on Ultra-fast Time-Scale

High-power laser radiation interacting with matter initiates many processes: starting with the linear absorption of laser radiation by free electrons or, because of the intense electrical field of the laser radiation $> 10^6$ V/m, free electrons are generated by non-linear ionization of atoms, like tunnel or multi-photon ionization (Section 3.3.1). The bound electrons, having partially absorbed the optical energy, are excited into another state. The manifold of processes are described in Section 3.3.2. As a consequence free electrons interact with each other, with atoms and with ions being in the solid, and heating (Section 3.3.2), melting, evaporation, and ionization result (Section 3.3.3). The evaporated and partially ionized matter together with electrons form a plasma. Some fundamental descriptions of plasmas are given in Sections 3.3.4 and 3.3.5. The interaction of plasmas with radiation and the dynamics of plasma are given in the Addendum C.1.

3.3.1
Absorption of Radiation in a Plasma

Laser radiation interacting with matter, such as dielectrics, at intensities above 10^{10} W/cm^2 induces non-harmonic oscillations of bound electrons generating free electrons. Different from non-linear optics presented above, here these free electrons partially absorb the optical energy resulting into an increased kinetic energy of the free electrons. These electrons can themselves initiate processes like impact ionization.

3.3.1.1 **Electromagnetic Waves in Plasmas**
The absorption and propagation of laser radiation in an inhomogeneous plasma have been described and the absorption can be calculated [91] knowing of the electromagnetic fields entering and exiting the plasma. Solving the Maxwell equations for stationary ions (ion plasma frequency ω_{pi}) with $\omega_L \geq \omega_{pe} \gg \omega_{pi}$ one gets the

electromagnetic fields

$$\vec{E}(\vec{r},t) = \vec{E}_0 \exp\left[i(\vec{k}\cdot\vec{r}-\omega t)\right] , \quad \vec{B}(\vec{r},t) = \vec{B}_0 \exp\left[i(\vec{k}\cdot\vec{r}-\omega t)\right] \tag{3.67}$$

and the dispersion relation, Eq. (3.77), without collisions between the electrons ($\nu_e = 0$)

$$k^2 = \frac{\omega_L^2 \varepsilon}{c^2} = \frac{1}{c^2}\left(\omega_L^2 - \omega_p^2\right) , \tag{3.68}$$

using the dielectric function

$$\varepsilon = 1 - \frac{\omega_p^2}{\omega_L^2} = 1 - \frac{n_e(\vec{r})}{n_{ec}} , \tag{3.69}$$

with n_{ec} the critical electron density, Eq. (3.83). These solutions for the electric field arise in an imaginary k for $\omega/\omega_{pe} < 1$ being equivalent to $n_{ec}/n_e < 1$. The electric field decays exponentially when passing through the plasma, and cannot propagate into the plasma with electron densities larger than the critical electron density n_{ec}.

Laser radiation passing a plasma is not always orthogonally incident, but can be obliquely incident. For example strong focusing onto a plasma with a large NA objective results in a divergence angle of $\approx 30°$ (see correlation experiments for the generation of X-rays Section 5.1.2, page 5.1.2.1). The linearly polarized radiation enters the plasma at the angle θ_0, the electric field is parallel to the x-axis for s-polarization and in y-z-plane for the p-polarization (Figure 3.6). Assuming also a linear density profile in one dimension the electron density

$$n_e(z) = n_{ec}\frac{z}{L} \tag{3.70}$$

can be calculated with the plasma scale length L, Eq. (3.85).

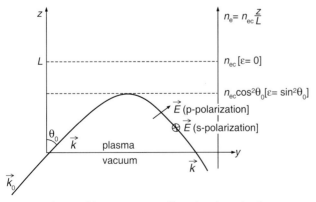

Fig. 3.6 Scheme of the propagation of linearly polarized radiation in the y-z-plane through a plasma [91].

As described in [91], by solving the Maxwell wave equation for the plasma parameters, for the s-polarization one obtains the absorptivity

$$A_{\text{s-pol}} = 1 - \frac{|E_{\text{out}}|^2}{|E_{\text{in}}|^2} = 1 - \exp\left(-\frac{32\nu_c L \cos^5 \theta_0}{15c}\right). \tag{3.71}$$

The solution of Maxwell's equations show that p-polarized radiation incident on a steeply rising plasma density at an angle different from zero has a singularity in the magnitude of the oscillating electric field in the plasma. This electric field resonantly drives an electron plasma wave. The absorptivity by resonance absorption is larger than by inverse Bremsstrahlung for the following variables: high plasma temperatures generated by high radiation intensities, large radiation wavelengths resulting in smaller critical electron density n_{ec} and small plasma scale length L.

For intensities above 10^{15} W/cm^2 (at 1 μm wavelength) at the appropriate polarization and angle of incidence up to 50% of the radiation is absorbed, generating mainly hot electrons, since only a minority of the plasma electrons acquire most of the absorbed energy in contrast to collision absorption. The latter heats all the electrons. For a linear density profile and p-polarized radiation one obtains [91]

$$\alpha_{\text{p-pol}} = 36\tau^2 \frac{[Ai(\tau)]^3}{|dAi(\tau)/d\tau|}, \tag{3.72}$$

where Ai is the asymptotic Airy function[24]

$$Ai(-\zeta) \underset{\zeta \to \infty}{\approx} \frac{1}{\sqrt{\pi}\zeta^{1/4}} \cos\left(\frac{2}{3}\zeta^{3/2} - \frac{\pi}{4}\right), \tag{3.73}$$

the plasma vacuum interface

$$\zeta = \left(\frac{\omega^2}{c^2 L}\right)^{1/3} (z - L), \tag{3.74}$$

and the $\tau = (\omega L/c)^{1/3} \sin \theta_0$. Different from s-polarized radiation, where the absorption coefficient decreases steadily with an increasing angle, p-polarized radiation exhibits a maximum absorption at about $25°$ (Figure 3.7).

3.3.1.2 Inverse Bremsstrahlung

Inverse Bremsstrahlung is the process by which an electron absorbs a photon while colliding with an ion or with another electron. By using the kinetic theory, and taking into account the distribution function of the electrons and ions, this process can be rigorously calculated. For simplicity's sake an approach with an infinite and homogeneous plasma with immobile ions (being comparable to a metal) without magnetic and electric fields will be described. The phase velocity of the electromagnetic field is much larger than the thermal velocity of the electrons, so the thermal

[24] Asymptotic Airy function $\lim_{\zeta \to \infty} Ai(\zeta)$,

$$Ai := \frac{1}{\pi}\int_{-\infty}^{\infty} \cos\left(\frac{t^3}{3} - \zeta t\right) dt$$

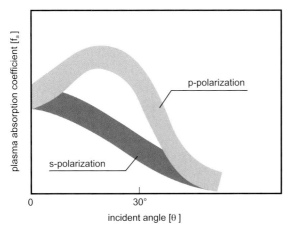

Fig. 3.7 Scheme of the plasma absorption coefficient for s- and p-polarized radiation as a function of the incident angle [91].

electrons can be neglected. The equation of motion for the electrons is given by

$$\frac{d\vec{v}}{dt} = -\frac{e\vec{E}}{m_e} - \frac{\vec{v}}{\tau_c}, \tag{3.75}$$

where the effective electron–ion collision time is $\tau_c = \nu_{eo}^{-1}$ defined by Eq. (3.114). The dispersion relation of radiation in this simple plasma model is calculated by using the Maxwell equations using the plasma frequency

$$\omega_p^2 = \frac{4\pi e^2 n_e}{m_e} \tag{3.76}$$

resulting in the dispersion relation for orthogonal and parallel field components

$$\vec{k} \cdot \vec{E} = 0: \quad k^2 = \frac{\omega_L}{c^2} - \frac{\omega_p^2 \omega_L}{c^2(\omega_L + i\nu_{ei})}, \tag{3.77}$$

$$\vec{k} \times \vec{E} = 0: \quad \omega^2 + i\nu_{ei}\omega = \omega_p^2. \tag{3.78}$$

Since the orthogonal field components couple directly to the laser radiation, so only Eqs. (3.77) and (3.78) will be considered further. For electron–ion collision frequencies which are much smaller than the radiation frequency $\nu_{ei} \ll \omega_L$, a Taylor expansion of the dispersion relation gives

$$k^2 \approx \frac{\omega_L^2}{c^2} \left(1 - \frac{\omega_p^2}{\omega_L^2} + \frac{i\nu_{ei}\omega_p^2}{\omega_L^3} \right). \tag{3.79}$$

The solution of Eq. (3.79) is obtained by expanding the square root for $\nu_{ei}/\omega_L \ll 1$ and using $\omega_L^2 - \omega_p^2 \gg (\nu_{ei}/\omega_L)\omega_p^2$

$$k \cong \pm \frac{\omega_L}{c} \left(1 - \frac{\omega_p^2}{\omega_L^2} \right)^{1/2} \left[1 + i \left(\frac{\nu_{ei}}{2\omega_L} \right) \left(\frac{\omega_p^2}{\omega_L^2} \right) \frac{1}{1 - \frac{\omega_p^2}{\omega_L^2}} \right]. \tag{3.80}$$

The change in intensity I passing through a slab of plasma in the z-direction is given by

$$\frac{dI}{dz} = -\kappa_{ib} I. \tag{3.81}$$

The spatial damping rate of optical energy by inverse Bremsstrahlung κ_{ib} is given by

$$\kappa_{ib} = 2 \operatorname{Im}(k) = \left(\frac{\nu_{ei}}{c}\right) \left(\frac{\omega_p^2}{\omega_L^2}\right) \left(1 - \frac{\omega_p^2}{\omega_L^2}\right)^{-1/2}. \tag{3.82}$$

When defining a critical electron density

$$n_{ce} = \frac{m_e \omega_L^2}{4\pi e^2} \tag{3.83}$$

one obtains for κ_{ib} with the electron–ion collision frequency Eq. (3.114)

$$\kappa_{ib} = \frac{\nu_{ei}(n_{ce})}{c} \left(\frac{n_e}{n_{ce}}\right)^2 \left(1 - \frac{n_e}{n_{ce}}\right)^{-1/2}, \tag{3.84}$$

with $\nu_{ei}(n_{ce})$ the collision frequency evaluated at the critical electron density n_{ei}. For a plasma slab of length L an absorptivity A can be calculated to

$$A = \frac{I_{in} - I_{out}}{I_{in}} = 1 - \exp\left(-\int_0^L \kappa_{ib} dz\right), \tag{3.85}$$

with I_{in} and I_{out} being the incoming and transmitted intensities, respectively. For a linear electron density profile of the plasma $n_e = n_{ce}(1 - z/L)$ with $0 \leq z \leq L$ one obtains the absorptivity

$$A = 1 - \exp\left(-\frac{32}{15} \frac{\nu_{ei}(n_{ce})L}{c}\right) \tag{3.86}$$

valid for small radiation intensities where the energy distribution of the electrons is not changed.

At intensities $I > 10^{15}\,\text{W}\,\text{cm}^{-2}$ the intense electric field of the laser radiation distorts the electron energy distribution, thus changing the electron–ion collision frequency ν_{ei}. The electron velocity in an electric field E_L of laser radiation is given by

$$v_E = \frac{eE_L}{m_e \omega_L}. \tag{3.87}$$

For oscillation energies of the electron comparable to the thermal electron energy v_{Te}, the effective electron velocity results in

$$v_{eff}^2 = v_{Te}^2 + v_E^2. \tag{3.88}$$

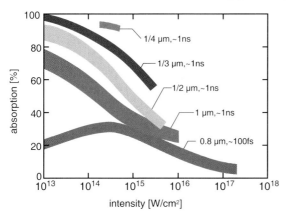

Fig. 3.8 Absorptivity at the surface of aluminum as a function of intensity for different wavelength and pulse duration [91].

At large intensities and for $v_E/v_{Te} > 1$ the spatial damping rate κ_{ib}, Eq. (3.84), can be approximated by

$$\kappa_{ib}^{hI} = \frac{\kappa_{ib}}{\left[1 + (v_E^2/v_{Te}^2)\right]^{3/2}}, \tag{3.89}$$

which for $v_E/v_{Te} < 1$ can be Taylor-expanded to

$$\kappa_{ib}^{hI} = \frac{\kappa_{ib}}{1 + \frac{3 v_E^2}{2 v_{Te}^2}}. \tag{3.90}$$

The experimental absorption coefficient for aluminum as a function of laser intensity and wavelength demonstrates a decrease in absorptivity for increasing intensity (Figure 3.8). Also the temporal dependence seems to become of secondary importance with increasing intensity, especially when using ultra-fast laser radiation. In this case, a Ti:sapphire laser radiation with pulse duration of 100 fs, a local increase in absorption due to resonance absorption is given at the intensity $I \approx 5 \cdot 10^{14} \, \text{W} \, \text{cm}^{-2}$.

3.3.2
Plasma Heating

Laser radiation interacting with matter exhibits different processes and can be described by three physical domains (Figure 3.9) [91]:

- **Absorption:** the density is $< 0.01 \, \text{g/cm}^3$ and the temperatures $T \approx 100 \, \text{eV}$,
- **Transport:** the density is between gas density $0.01 \, \text{g/cm}^3$ and solid density ϱ_0 with temperatures between $T \approx 30 \, \text{eV}$ and $100 \, \text{eV}$, and
- **Compression:** the density is between solid density ϱ_0 and $10\varrho_0$ with temperatures between $T \approx 1 \, \text{eV}$ and $30 \, \text{eV}$.

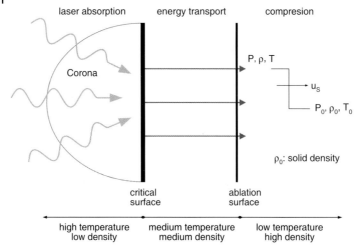

Fig. 3.9 Scheme of laser radiation-plasma-matter interaction.

Laser radiation with pulse durations $t_p \geq 1$ ns generates within the absorption domain a corona[25] and is subsequently absorbed by the electrons of the corona [91]. The blowout velocity u_{bo} of the plasma is about the speed of sound c_T at the critical plasma density

$$u_{bo} \approx c_T = \left(\frac{Zk_B T_e}{m_i} \right)^{1/2}. \tag{3.91}$$

The size of the corona becomes about c_T/t_p. In the case of ultra-fast laser radiation with pulse durations $t_p < 1$ ns, the corona extension is small and can often be neglected. A measure of the absorption extension within a corona is given by the skin depth

$$\delta = \frac{c}{\omega_p}. \tag{3.92}$$

Depending on the intensity of the laser radiation, two classes of electrons are generated by ablation: "cold electrons" with a temperature T_e^c on the order of 1 eV and "hot electrons" with temperatures $T_e^H > 10$ keV.

- At small intensities $< 10^{12}$ W/cm^2 electrons with energies below 1 eV are generated by gathering energy from the electromagnetic field through collisional absorption (inverse Bremsstrahlung). The electrons and the ions feature before irradiation the same temperatures $T_e = T_i$, being in LTE. The cross-section for collisional absorption by inverse Bremsstrahlung scales by

$$\sigma_{coll} \propto \frac{1}{T^{1/2}}. \tag{3.93}$$

25) The plasma from the critical density outwards
(towards the laser source) is defined as the
corona.

With increasing temperature of the electrons by collisional absorption the cross-section σ_{coll} decreases, resulting in no more heating.

- At larger intensities the electrons from the volume with density n_e^{crit} are displaced by interacting with the electromagnetic field into regions with smaller densities. There the electrons gather large energy above 1 keV by collisional absorption (resonance absorption). Because of resonance absorption at the plasma with the critical electron density n_e^{crit}, the temperature of the electrons scale by the intensity

$$T_e \propto I^{1/2}.\tag{3.94}$$

Assuming a temporal Gaussian intensity distribution, small intensities are reached well before the maximum of the laser pulse. This results in the generation of low-energy electrons by collisional absorption. At large intensities of the same pulse $I_L > 10^{12}$ W/cm^2 high-energy electrons are generated by resonant absorption. A local thermodynamic equilibrium between the hot and cold electrons cannot be defined, because of the thermal diffusion of the "cold electrons" carrying energy out from the absorption region. The "hot electrons" transport the energy in beam direction behind the thermal conduction front, causing electron pre-heating. Because of the pre-heating at intensities $> 10^{14}$ W/cm^2 (for example, for the wavelength $\lambda \approx 1\,\mu$m) the energy transport by thermal conduction is disturbed. The surface of the matter is exposed to different forces resulting from the pressures applied to it.

- The radiation pressure

$$P_L = \frac{I_L}{c}(1 + R),\tag{3.95}$$

is proportional to the intensity of laser radiation and depends on the reflectivity, becoming dominant at intensities $I_L > 10^{15}$ W cm^{-2}. This pressure rises above 1 Gbar for intensities $I_L \geq 3 \cdot 10^{18}$ W cm^{-2} with the effect of steepening the density gradient of the plasma close to the critical density.

- The electron pressure generated by the "hot electrons" P_H, the "cold electrons" P_e and the ions P_i are associated with the temperatures T_e^H, T_e^c, T_i. Assuming an ideal gas for the electrons and the ions, one obtains

$$P_e^c = n_e k_B T_e^c,$$

$$P_e^H = n_H k_B T_e^H,\tag{3.96}$$

$$P_i = n_i k_B T_i.$$

- The resulting recoil pressure P_a is associated with the flow of heated vapor and plasma from the solid surface. The ablation pressure drives the shock wave into the solid and compresses the solid. Because of a inhibition of energy transport by the "hot electrons" the ablation pressure is reduced. For a high-density cold plasma located on a low-density hot plasma, hydrodynamic instabilities may occur (Addendum C.5).

3.3.3
Ionization of a Plasma

The discussion of plasma generated by intense ultra-fast laser radiation is closely related to the work of Eliezer [91]. Ionization is the process where a neutral atom becomes an ion. Or more generally, an atom (or ion, denoted by A_j) that has lost j ($j = 0, 1, 2 \ldots$) electrons is transformed to an atom that has lost $j + 1$ electrons (A_{j+1}). In thermal equilibrium with N particles per unit mass of the ions, A_j and A_{j+1}, the following chemical equations are valid

$$A_j \longleftrightarrow A_{j+1} + e^-, \quad j = 0, 1, 2, \ldots \tag{3.97}$$

$$\delta N_j = -\delta N_{j+1} = \delta N_e . \tag{3.98}$$

The thermodynamic state can be described by the free energy F. In a thermodynamic equilibrium the free energy $F(T, V, N)$ (variables for F: T temperature, $V = \varrho^{-1}$ volume as inverse of the density, and N number of particles) is minimized:

$$F(T, V, N) = -\sum_j N_j k_B T \ln \frac{e Q_j}{N_j} - N_e k_B T \ln \frac{e Q_e}{N_e} \tag{3.99}$$

$$\delta F = \sum_j \frac{\delta F}{\delta N_j} \delta N_j + \frac{\delta F}{\delta N_e} \delta N_e = 0 . \tag{3.100}$$

Hereby, Q_j and Q_e are the partition functions for ions and electrons, defined as

$$Q = \sum_i \exp \left(-\frac{\varepsilon_i}{k_B T} \right) , \tag{3.101}$$

k_B the Boltzmann constant and ε_i the energy eigenstates of the Hamiltonian describing the plasma system. Solving Eqs. (3.98), (3.99) and (3.100) results in

$$\frac{N_{j+1} N_e}{N_j} = \frac{Q_{j+1} Q_e}{Q_j} , \quad j = 0, 1, 2, \ldots \tag{3.102}$$

implying the Saha equation

$$\frac{n_{j+1} n_e}{n_j} = \frac{2 U_{j+1}}{U_j} \left(\frac{2 \pi m_e k_B T}{h^2} \right)^{3/2} \exp \left(-\frac{I_j - \Delta I_j}{k_B T} \right) , \quad j = 1, 2, \ldots, (Z - 1) \tag{3.103}$$

with n_{j+1} and n_j the densities ($n = N/V$) of the $(j + 1)$-th and j-th ionization state, n_e the electron density, U_{j+1} and U_j the internal parts of the ionic partition function, m_e the electron mass, h the Planck's constant, I_j the ionization energy of the ground state and $\Delta I_j = \varepsilon_{j+1,0} - \varepsilon_{j,0}$ the reduction of the ionization potential due to local electrostatic fields in the plasma.

The Saha equation is valid for plasmas in a local thermodynamic equilibrium (LTE). Dynamical properties like electron and ion velocities, population partition, and ionization state densities follow the Boltzmann distribution for LTE. LTE for thermodynamic equilibria are valid except for Planck's radiation law. LTE may be valid even when there are temperature gradients in the plasma. For high-density plasmas, where the frequent collisions between electrons and ions or electrons between themselves produce equilibrium, LTE is satisfied. In plasmas energy can diffuse. Then the plasma is not necessary in thermal equilibrium with the plasma particles.

3.3.3.1 Cross-Section and Collision Frequency

The cross-section σ_{ab} defines the collision probability

$$dF = -\sigma_{ab} F_a n_b dl \tag{3.104}$$

for a particle beam with flux F_a to collide with other steady state particles (density n_b) of thickness dl. This cross-section also defines the absorption coefficient

$$\mu_a = \sum_b n_b \sigma_{ab} = \frac{1}{l_a}, \tag{3.105}$$

with the sum taken over all the species in the plasma and the attenuation, also called mean free path l_a.

The collision frequency

$$\nu_{ab} = n_b \sigma_{ab} v_a = \frac{v_a}{l_a} \tag{3.106}$$

between a particle a with velocity $\vec{v}_a$ and a medium with particles b is defined as the number of collisions per second of the particle. Coulomb forces influences the trajectories of charged particles in ionized plasma. The Rutherford differential scattering cross section

$$\frac{d\sigma_{ei}}{d\Omega} = \left(\frac{1}{4}\right)\left(\frac{Ze^2}{m_e v^2}\right)^2 \frac{1}{\sin^4(\theta/2)} \tag{3.107}$$

describes the collision between an electron and an ion at rest with charge Ze, with m_e and v representing the electron mass and velocity, θ the scattering angle and $d\Omega$ the differential solid angle. The scattering angle θ is related to the impact parameter by

$$b \tan\frac{\theta}{2} = \frac{Ze^2}{m_e v^2}, \tag{3.108}$$

defined asymptotically before the Coulomb force becomes effective (Figure 3.10).

The total cross section

$$\sigma_{ei} = \int_{-\infty}^{\infty} d\Omega \frac{d\sigma_{ei}}{d\Omega} = \frac{\pi}{2}\left(\frac{Ze^2}{m_e v^2}\right)^2 \int_0^\pi d\theta \frac{\sin\theta}{\sin^4(\theta/2)} \tag{3.109}$$

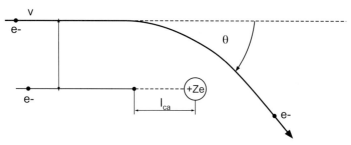

Fig. 3.10 Scheme of the trajectory of an electron colliding with a positive ion of charge Ze.

is given and diverges for $\theta = 0$, or being equivalent for $b \to \infty$. The long-range interactions are screened by the charged particles in the vicinity so that there is an effective b_{max} instead of $b \to \infty$; b_{max} is taken as the Debye length (Addendum C.2) and $b = 0$ is replaced by the closest approach l_{ca} given by

$$\frac{Ze^2}{l_{ca}} = k_B T_e .$$

(3.110)

An estimation for the electron ion collision frequency can be made with the total cross section

$$\sigma_{ei} \approx \pi l_{ca}^2 \propto T_e^{-2} .$$

(3.111)

Assuming a Maxwell distribution, velocity distribution of electrons at LTE is given and one gets the thermal velocity

$$v_T^2 = \frac{2k_B T}{m_e} .$$

(3.112)

This equation combined with Eqs. (3.106) and (3.111) and $v_a = v_T$ results in an estimate for the electron–ion collision frequency

$$\nu_{ei} \approx \frac{\sqrt{2\pi} Z^2 e^4 n_i}{\sqrt{m_e} (k_B T_e)^{3/2}} .$$

(3.113)

An accurate calculation of ν_{ei} [91] results in

$$\nu_{ei} = \frac{4\sqrt{2\pi} Z^2 e^4 n_i \ln \Lambda}{3\sqrt{m_e} (k_B T_e)^{3/2}} ,$$

(3.114)

with the collision corridor

$$\Lambda = \frac{b_{max}}{b_{min}} .$$

(3.115)

3.3.4

Electron Transitions and Energy Transport

An electron as a part of a plasma can transit from an initial state to a final state by emitting or absorbing radiation. As shown in Figure 3.11 different transitions are possible:

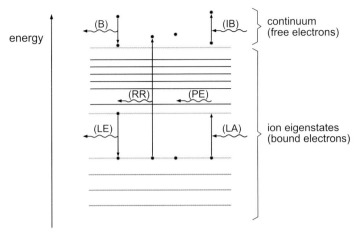

Fig. 3.11 Electron transitions in an ion or atom associated with radiation emission or absorption [91].

- **Bound–bound (bb):** A photon is emitted by changing the quantum energetic level (line emission) $i_1 \rightarrow i_2 + \gamma$ or a photon is absorbed (line absorption) $\gamma + i_2 \rightarrow i_1$,
- **Bound–free (bf):** A photon is emitted when a free electron is caught by an ion (radiative recombination) $e^- + i_1 \rightarrow i_2 + \gamma$ or the inverse process, a photon is absorbed by an ion and a free electron is generated (photoelectric effect) $\gamma + i_2 \rightarrow i_1 + e^-$, and
- **Free–free (ff):** An electron colliding with an ion emits radiation as a photon (Bremsstrahlung) $e^- + i \rightarrow e^- + i + \gamma$ whereas the inverse process plays a major role in laser radiation absorption (inverse Bremsstrahlung) $e^- + i + \gamma \rightarrow i + e^-$.

A photon as a particle is characterized by its energy E_ν, wavelength λ and momentum p_ν, which are correlated to the frequency ν and speed of light c by

$$\lambda = \frac{c}{\nu}, \quad E_\nu = h\nu, \quad \text{and} \quad p_\nu = \frac{h\nu}{c}.$$

The refractive index of a plasma is given by

$$n_R = \left(1 - \frac{\nu_{pe}^2}{\nu^2} \right)^{1/2}, \tag{3.116}$$

where ν_{pe} defines the electron oscillation frequency and the speed of light in a plasma $c_p = c/n_R$.

In order to write the transport equation one needs to define some parameters, like the total emission described by the emissivity, the induced emission, and the opacity.

3.3.4.1 The Total Emission

The Emissivity j_ν describes the spontaneous emission of the medium, and depends on the atoms of the medium, the degree of ionization and the temperature. The spontaneous emission in energy per volume and time is given by

$$j_{se} = j_\nu d\nu d\Omega .$$ (3.117)

The spectral radiation intensity I_ν is defined as the radiation energy per frequency between ν and $\nu + d\nu$ crossing a unit area per time in direction Ω within the solid angle $d\Omega$.

The induced emission j_{ie} is given by

$$j_{ie} = j_\nu \left(\frac{c^2 I_\nu}{2h\nu^3} \right) d\nu d\Omega .$$ (3.118)

The term in the parentheses defines the number of photons in the same phase space cell as the emitted photon.

3.3.4.2 The Opacity

Absorption and scattering can be written as

$$j_a = \kappa_\nu I_\nu d\nu d\Omega$$ (3.119)

with the opacity

$$\kappa_\nu = \frac{1}{l_\nu} = \sum_j n_j \sigma_{\nu j} ,$$ (3.120)

and the mean free path l_ν, the density n_j of the particle j and the appropriate cross-section for absorption or scattering $\sigma_{\nu j}$.

By combination of the Eqs. (3.117), (3.118) and (3.119) given by the total derivative of the spectral radiation intensity with respect to time, one obtains the transport equation

$$\frac{dI_\nu}{dt} = \frac{1}{c} \left(\frac{\partial I_\nu}{\partial t} + c\vec{\Omega} \cdot \nabla I_\nu \right) = j_{se} + j_{ie} - j_a$$ (3.121)

$$= j_\nu \left(1 + \frac{c^2 I_\nu}{2h\nu^3} \right) - \kappa_\nu I_\nu .$$ (3.122)

In thermal equilibrium the ratio of the spontaneous emission j_ν to the absorption k_ν is a universal function of the frequency and temperature and is given by

$$\frac{j_\nu}{k_\nu} = \frac{2h\nu^3}{c^2} \exp\left(-\frac{h\nu}{k_B T} \right) = I_{\nu p} \left[1 - \exp\left(-\frac{h\nu}{k_B T} \right) \right] ,$$ (3.123)

$$I_{\nu p} = \frac{2h\nu^3}{c^2} \left[\exp\left(\frac{h\nu}{k_B T} \right) - 1 \right]^{-1}$$ (3.124)

$$j_\nu = \kappa_\nu \left[1 - \exp \left(-\frac{h\nu}{k_B T} \right) \right] I_{\nu p} \,. \tag{3.125}$$

Equation (3.125), known as Kirchhoff's law, describes the balance between emission and absorption. By substituting j_ν in the second term of Eq. (3.122) by Eqs. (3.125) and (3.124) one becomes

$$j_\nu \frac{c^2 I_\nu}{2h\nu^3} = -\kappa_\nu I_\nu \,. \tag{3.126}$$

One gets finally for the transport equation in thermal equilibrium

$$\frac{\partial I_\nu}{c\partial t} + \vec{\Omega} \cdot \nabla I_\nu = \kappa'_\nu \left(I_{\nu p} - I_\nu \right) \,, \tag{3.127}$$

with $\kappa'_\nu = \kappa \left[1 - \exp \left(-\frac{h\nu}{k_B T} \right) \right]$. This equation can be applied to the plasma in local thermal equilibrium (LTE).

3.3.5
Dielectric Function of a Plasma

High-power laser radiation interacting with a plasma is reflected, deflected, and absorbed by the plasma. The optical properties of a plasma are summarized by the dielectric constant. A plasma can be described as a dielectric medium with the scalar dielectric function. Assuming the electrical charges described by

$$\varrho_e = -e n_e + q n_0 \quad \vec{J}_e = -e n_e \vec{v}_e \,, \tag{3.128}$$

the ions n_0 are stationary and serve as a charge-neutralizing background to the electron motion. For a monochromatic electromagnetic field

$$\vec{E}(\vec{r}, t) = \vec{E}(\vec{r}) \exp(-i\omega t) \,, \quad \vec{B}(\vec{r}, t) = \vec{B}(\vec{r}) \exp(-i\omega t) \tag{3.129}$$

the electron velocity $\vec{v}_e$ is calculated by using Newton's law

$$\frac{\partial \vec{v}_e}{\partial t} + \nu_e \vec{v}_e = -\frac{e}{m_e} \vec{E}(\vec{r}) \exp(-i\omega t) \,, \tag{3.130}$$

where ν_e defines the electron collision frequency. The solution of Eq. (3.130) results in

$$\vec{v}_e(\vec{r}, t) = \frac{-ie\vec{E}(\vec{r}, t)}{m_e(\omega + i\nu_e)} \,. \tag{3.131}$$

Substituting Eq. (3.131) into Eq. (3.128), the current

$$\vec{J}_e(\vec{r}, t) = \sigma_E \vec{E}(\vec{r}, t) \,, \tag{3.132}$$

is described by applying the plasma frequency, Eq. (C33), with

$$\sigma_E = \frac{i\omega_{pe}^2}{4\pi(\omega + i\nu_e)}$$

(3.133)

representing the electrical conductivity. Inserting Eq. (3.132) into the Maxwell equations

$$\nabla \times \vec{B} = \frac{1}{c}\frac{\partial \vec{E}}{\partial t} + \frac{4\pi}{c}\vec{J}_e$$

(3.134)

and assuming linear functions using $\partial_t = -i\omega$, one gets

$$\nabla \times \vec{B} = \frac{1}{c}\frac{\partial(\varepsilon\vec{E})}{\partial t} .$$

(3.135)

In this way the dielectric function given by Eq. (3.77) can be rewritten as

$$\varepsilon = 1 - \frac{\omega_{pe}^2}{\omega(\omega + i\nu_e)} = 1 + i\frac{4\pi\sigma_E}{\omega}$$

(3.136)

and is used. For a spatial dependence of the electromagnetic field $\propto \exp(i\vec{k}\cdot\vec{r})$ from the Maxwell equations one obtains the dispersion relation for the electrical field in a plasma

$$k^2 = \frac{\omega^2\varepsilon}{c^2} \qquad \Leftrightarrow \qquad \omega^2 = \omega_{pe}^2 + k^2c^2.$$

(3.137)

The refractive index of a plasma is then given by

$$n_p = \sqrt{\varepsilon} = n_R + in_I .$$

(3.138)

3.3.6
Ponderomotive Force

In physics, a ponderomotive force is a nonlinear force that a charged particle experiences in a rapidly oscillating, inhomogeneous electric or electromagnetic field [127]. A first description of the ponderomotive force has been given by Kelvin ($\approx$ 1850) and Helmholtz (1881). Both attempts failed, and in the 1950s Landau and Lipschitz achieved the correct expression [128, 129]. The ponderomotive force f_p is given by

$$\vec{f}_p = -\frac{\omega_p^2}{16\pi\omega_L^2}\nabla\vec{E}_S^2 ,$$

(3.139)

where ω_p is the plasma frequency, Eq. (C33), ω_L is the oscillation frequency of the field, and $\vec{E}_S$ is the space-dependent electric field

$$\vec{E} = \vec{E}_S(\vec{r}) \cos \omega_L t .$$

(3.140)

The Eq. (3.140) describes an electron in an inhomogeneous oscillating field, not only oscillating at the frequency of ω_L, but also drifting toward the weak field area.

The ponderomotive force is involved in many physical phenomena of laser physics in plasmas (references in [91]):

- momentum transfer to a target,
- self-focusing and filamentation of the laser beam,
- plasma density distribution change,
- parametric instabilities,
- second harmonic generation, and
- magnetic field generation.

The mechanism of the ponderomotive force can be understood by considering the motion of the charge in an oscillating electric field. In the case of a homogeneous field, the charge returns to its initial position after one cycle of oscillation. In contrast, in the case of an inhomogeneous field, the position that the charge reaches after one cycle of oscillation shifts toward the lower field amplitude area. Since the force imposed onto the charge at the turning point with a higher field amplitude is larger than that imposed at the turning point with a lower field amplitude, the result is a net force driving the charge towards the weak field area.

4
Fundamentals of Pump and Probe

The important parameters of ultra-fast laser radiation, like pulse duration and coherence, have to be determined in order to carry out a pump and probe experiment.
The probe radiation has to be prepared in order to give reliable information on the
investigated process.

Ultra-fast laser radiation used to probe a state of matter has to be guided from
the source by reflective optics to the sample. In this way some properties of the
radiation, like pulse duration, spectral bandwidth, and coherence can be changed
(Section 4.1). Ultra-fast laser radiation has to be handled with more caution than
monochromatic laser radiation with large pulse durations ($t_p > 1$ ns) when applied
to transmitting optical elements like lenses, plates and objectives.

The state of matter can be prepared by the probe radiation, meaning that the radiation itself has to be prepared and characterized (Section 4.2). The probe laser radiation is often used to illuminate the interaction volume and the surroundings. The
basics of diffraction theory of optical radiation and of optical microscopy are given
in Section 4.3. In this way differences in the resolution of imaging systems with
non-coherent and coherent radiation are given. For pump and probe experiments
the probe radiation has to be temporally delayed relative to the pump radiation.
Different approaches are described in Section 4.4.

4.1
Ultra-fast Laser Radiation

The propagation of ultra-fast laser radiation adopted in pump and probe radiation
is described in Section 4.1.1. Ultra-fast laser radiation can be described, in terms of
common diffraction-limited monochromatic laser radiation in the technical field,
by the beam parameters pulse duration, wavelength, spectral band width, and polarization. Physically ultra-fast laser radiation adopt a spatial and temporal change
passing a dielectric material:

- **Spatial:** the phase and the pulse front positions of the radiation are unequal
  due to dispersion, because then the velocities of the pulse and phase fronts
  are different (Section 4.1.2)

Ultra-fast Material Metrology. Alexander Horn
Copyright © 2009 WILEY-VCH Verlag GmbH & Co. KGaA, Weinheim
ISBN: 978-3-527-40887-0

- **Temporal:** the spectral components of the radiation are delayed, so-called chirped.

In addition to the temporal-spatial distribution of the laser radiation, the chromatic coherence has to be taken into account.

4.1.1
Beam Propagation

For the beam guiding of ultra-fast laser radiation with peak intensities $> 10^{12}$ W/cm^2 a special treatment of beam propagation due to non-linear effects in condensed media is necessary (Section 3.1). This treatment can be avoided by

- decreasing the intensity by increasing the beam diameter or
- increasing the threshold intensity for self-focusing by changing the propagation medium to one with smaller non-linear coefficients or
- canceling self-focusing by moving the experiment into a vacuum when the intensity exceeds $I > 10^{12}$ W/cm^2.

The last choice is the best one, although the most complex and most expensive. Due to vibrations of the vacuum apparatus conveyed on the experiment, additional efforts have to be made to get a vibration-free set-up, especially for sub-micrometer investigations and applications.

4.1.2
Dispersion

In order to achieve large intensities ultra-fast laser radiation has to be concentrated on one spot by focusing. Passing optical elements, laser radiation with large spectral bandwidth like ultra-fast laser radiation suffers on pulse duration broadening, because the dispersion of the refractive index is wavelength-dependent.

As described in [72], the Maxwell wave Eq. (3.5) is solved using Eq. (3.6) by the field amplitude of a plane wave passing a glass plate of thickness L which results in an electrical field

$$\tilde{E}_2(\omega) = \tilde{E}_1(\omega) e^{\left(-iL[k(\omega)-\omega/c]\right)} \tag{4.1}$$

$$\approx \tilde{E}_1(\omega) e^{\left(-i\left[\left(k_l-\frac{\omega}{c}\right)L+k_l'L(\omega-\omega_l)+\frac{k_l''L}{2}(\omega-\omega_l)^2\right]\right)}, \tag{4.2}$$

with k being Taylor expanded for the carrier frequency ω_l, Eq. (3.7), $(k_l')^{-1} = v_g$ being the group velocity, and k_l'' the group velocity dispersion (GVD), Eq. (3.9). A glass plate of thickness L introduces, neglecting GVD, a temporal delay

$$\Delta\tau = \frac{L}{c}\left(\underbrace{(n-1)}_{\text{first term}} - \underbrace{\lambda_l\left[\frac{dn}{d\lambda}\right]_{\lambda_l}}_{\text{second term}}\right), \tag{4.3}$$

resulting from the difference of the length of the optical path with and without the glass (first term) and by the group velocity v_g (second term), defined by

$$v_g^{-1} = \frac{1}{c}\left(n - \lambda\frac{dn}{d\lambda}\right) .$$

(4.4)

A third term of the Taylor expansion of $k(\omega)$ resulting from the GVD (Section 3.1.1) would account for a deformation of the phase. The group velocity delay $\frac{L}{c}\lambda_l\left[\frac{dn}{d\lambda}\right]_{\lambda_l}$ describes a slipping of the pulse envelope with respect to the waves, whereas the GVD causes different parts of the pulse to travel at different velocities resulting in pulse deformation. As a consequence the phase front can be tilted in the focal regime with respect to the pulse front.

4.1.3
Coherence

Coherence is one accentuated property of laser radiation. Laser radiation can coherently interact with matter. The coherent interaction of radiation with matter implies a smaller pulse duration of the radiation than the phase memory of the excited medium. Coherent interaction of laser radiation with matter becomes important in pump and probe metrology for ultra-fast laser radiation with pulse durations $\ll 50\,\text{fs}$ [72]; however, available sources today are not advanced enough for industrial applications. Therefore, the coherent interaction will not be considered.

On the other hand, laser radiation interacts coherently with radiation. Coherence is adopted in pump and probe metrology for time-resolved visualization of phase changes. The coherence of the electrical fields E_1 and E_2 is quantified by the cross-correlation function

$$\langle E_1 E_2^* \rangle = \lim_{T\to\infty} \frac{1}{2T}\int_{-\infty}^{\infty} E_1(t)E_2^*(t-\tau)d\tau .$$

(4.5)

The cross-correlation describes the ability to predict the value of the second field by knowing the value of the first. The second field need not be different from the first one. In this case the cross-correlation becomes an autocorrelation function. This cross-correlation function is called the mutual coherence function in statistical optics. Optical coherence of ultra-fast laser radiation exists as temporal, spatial and spectral coherence.

- **Temporal coherence** is defined as the correlation between the value of an electrical field at two different times, separated by delay τ

$$\Gamma_{11}(\tau) = \langle E_1(t-\tau)E_1^*(t)\rangle .$$

(4.6)

  Ultra-fast laser radiation exhibits a temporal coherence duration comparable to the pulse duration t_p.
- **Spatial coherence** describes the averaged over time interference at two points in space, x_1 and x_2, of an electrical fields. The spatial coherence is

the cross-correlation of two points in space for an electric field

$$\Gamma_{12}(\tau) = \langle E_1(t - \tau) E_2^*(t) \rangle \tag{4.7}$$

The distance between two points which show the significant interference is called the spatial coherence length l_c. Laser radiation in general exhibits absolute spatial coherence over the whole beam area.

Consider cross-correlation function of the electrical field scalars $E_1(t)$ and $E_2(t)$ that fluctuates with time t at different two points $\vec{r}_1$ and $\vec{r}_2$ [62]

$$\Gamma_{12}(\tau) = \langle E_1^*(t) E_2(t + \tau) \rangle . \tag{4.8}$$

The mutual coherence function with $\tau = 0$ is named as the mutual intensity, given by

$$J_{12} = \Gamma_{12}(0) . \tag{4.9}$$

The complex degree of coherence is defined as the normalized mutual coherence

$$\gamma_{12}(\tau) = \frac{\Gamma_{12}(\tau)}{\sqrt{I_1 I_2}} \tag{4.10}$$

with the self-coherence

$$I_j = \Gamma_{jj}(0) , \quad (j = 1, 2) . \tag{4.11}$$

The complex degree of coherence γ_{12}, which is defined by Eq. (4.10), is related to the visibility of the spatial interference fringes that appear in the two-beam interference of radiation. The visibility defines the modulation depth of the intensity ranging from $\mathcal{V} = 1$: complete coherence to $\mathcal{V} = 0$: no coherence. The relationship between the complex degree of coherence and the visibility

$$\mathcal{V} = \frac{I_{max}(\vec{r}) - I_{min}(\vec{r})}{I_{max}(\vec{r}) + I_{min}(\vec{r})} = |\gamma_{12}(\tau)| \tag{4.12}$$

at a point $\vec{r}$, where $I_{max}(\vec{r})$ and $I_{min}(\vec{r})$ are the maximum and the minimum intensities in the radiation field. The complex degree of coherence describes the spatial correlation of the quasi-monochromatic electric fields in the space-time domain. Using ultra-fast laser radiation the visibility is spatially and temporally limited, so the apparatus has accordingly been adapted in order to get phase information at maximal visibility.

- **Spectral coherence** has to be taken into account for the spatial correlation of spectrally broad laser radiation. One effect which accounts for the spectral coherence having a central role is the correlation-induced spectral change, known as the Wolf effect [130]. Electric fields of different frequencies can interfere to form a pulse if they have a fixed relative phase relationship. Conversely, if waves of different frequencies are not coherent, then, when combined, they create a wave that is continuous in time (for example white light or white noise). A measure for spectral coherence is the pulse duration-bandwidth product (PBP). Laser radiation with pulse durations $\tau_p \leq 200$ fs and intensities $I > 10^{10}$ W/cm^2 interacting with dielectrics exhibits non-neglecting spectral bandwidth. The pulse duration τ_p and the spectral bandwidth $\Delta\omega_p$ of the radiation are interconnected by the pulse duration-bandwidth product

$$\Delta\omega_p t_p = 2\pi\Delta\nu_p t_p \geq 2\pi c_B , \tag{4.13}$$

c_B being a numerical factor with values $c_B = 0.441$ for Gaussian, $c_B = 0.315$ for hyperbolic secant (sech), and $c_B = 0.142$ for Lorentz temporal profiles [72]. If the phase depends linearly on the frequency (in other words $\phi(\omega) \propto \omega$) the pulse will have the minimum time duration for its bandwidth (as a transform-limited pulse), otherwise it is chirped by $b \cong \phi(\omega) \propto \omega$. For laser radiation with pulse durations $\tau_p < 200$ fs, the PBP reveals a non-monochromatic laser radiation with a spectral bandwidth > 10 nm. As a consequence of dispersion in dielectrics, the pulse duration of the bandwidth-limited laser radiation increases. The important parameter pulse duration for pump and probe metrology is connected to the spectral bandwidth as given by Eq. (4.13). The spectral bandwidth of the radiation can hinder the experiments, or at least increase the efforts, because the chirp of the radiation has to be detected.

4.2
Preparation of States and Conditions for Probe Beams

Irradiation of matter by ultra-fast laser radiation induces physical and chemical changes of the substrate. A pump and probe experiment enables the detection of changes induced by the pump beam, like the reflectivity change by laser-induced melting. The pump beam excites matter by heating and melting. In general a pump and probe experiment applies two or more beams:

- one beam, the pump beam, for the change of some kind of physical, chemical or biological property.
- the second one, the probe beam, for the detection of the property changes.

The change of the physical system is called preparation of the system and is described in Section 4.2.1. Techniques to change the properties of states by temporal shaping of ultra-fast laser radiation is given in Section 4.2.2. The techniques adopted to measure the prepared state are given is Section 4.2.3 and those techniques adopted to measure the probe radiation conditions are given in Section 4.2.4.

4.2.1
Preparation of States

Using the probe beam the property of the matter under investigation, here the reflectivity, is investigated by detecting changes of the probe beam parameters:

- energy,
- wavelength,
- chirp,
- polarization,
- pulse duration, and
- intensity distribution.

A state of matter is described by the properties of matter, like the density, the reflectivity, the electrical conductivity, and so on. The pump beam should change only one property and the probe beam must be detected without the change of these properties. The number of changed properties depends on the beam parameters of the probe radiation. In general, it is necessary to know all involved properties of matter with infinite precision.

A perspective to make a measurement applicable is given by preparing the experiment in an adequate way, changing only one process-relevant property, with negligible change of others. This is the crucial part of preparing the experiment. The preparation of the matter is achieved by selectively using and varying the process parameters of the radiation. Orthogonal polarized laser radiation, for example, exhibits a selective excitation of matter with the pump radiation. The exciting matter is probed independently from the pumping radiation by using polarizer as analyzing filters.

4.2.2
Preparation of States by Spectro-Temporal Shaping

The preparation of a state by the probe radiation can be given by temporal shaping of the probe radiation[26]. In the simplest way, the original pulse distribution defined by the pulse duration drives the process (Figure 4.1). The temporal shape of the probe radiation is adapted in order to initiate a definite reaction. Pulse shaping can be obtained by

- spatial light modulation using liquid crystal display (LCD), or
- acousto-optic modulation using an AOPDF (acousto-optic programmable dispersive filter)

On time scales < 1 ns no devices exist for temporal modulation of radiation, for which reason the pulse shaper modulates the spectral distribution. Using evolu-

---

26) contents taken partly from a lecture of
the Georgia Center for Ultrafast Optics
http://www.physics.gatech.edu/gcuo/

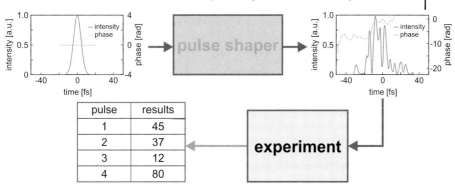

Fig. 4.1 Scheme of tailoring a pulse shape (intensity and phase) in a specific controlled manner by a pulse shaper resulting in a adapted intensity and phase distribution[26].

tionary or genetic algorithms, chemical reactions are optimized by a self-learning process (Figure 4.1).

A spatial light modulator uses an all-optical Fourier transform to achieve a temporal modulation of the laser radiation by using a zero-dispersion stretcher. The grating disperses the radiation setting a correlation between the wavelength of the radiation and the diffraction angle. The first lens maps angle (hence wavelength) to position and the second lens and grating undo the spatiotemporal distortions (Figure 4.2a). A phase mask made from liquid crystals selectively delays colors and controls the amplitude and phase of the laser pulse. The two masks or "spatial light modulators" together can yield any desired pulse. Liquid crystals orient along an applied DC electrical field and induce a phase delay (or birefringence) that depends on an applied voltage. Liquid crystals can yield both phase and amplitude masks.

The acousto-optic programmable dispersive filter works without a zero dispersion stretcher and hence without spatiotemporal pulse distortions. It generates an acoustic wave along the beam in a birefringent crystal (Figure 4.2b). The input po-

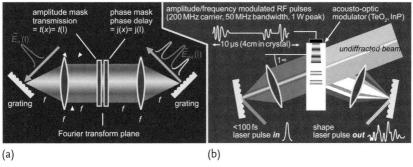

(a)   (b)

Fig. 4.2 Scheme of a spatial modulator (a) and principle acousto-optic programmable dispersive filter (b).

larization is diffracted by the sound wave. The frequency of the rotated polarization depends on the acoustic-wave frequency. Its relative delay at the crystal exit depends on the relative group velocities of the two polarizations. The extra phase delay seen by each wavelength depends on the penetration depth of the acoustic wave into the crystal on that wavelength and on the ordinary and extraordinary refractive indices. The strength of the acoustic wave at each wavelength determines the amplitude of the output wave at that wavelength.

4.2.3
Measurement of States

The excited state under investigation has to be detected and is sampled by the probe radiation. The choice of state depends on the physical process under investigation and the effect of the probe radiation on it. A state of matter can be interrogated differently:

- Sampling of a state is achieved by optical measuring of one beam parameter. The state under investigation should not be changed by the probe radiation. The analogy to an idealized quantum mechanical experiment is here given. For example, using the second harmonic of the pump laser radiation the reflectivity as a function of the angle of incidence and the polarization can be analyzed.
- Excitation of the initial state by the probe radiation into a definite exited state. This state relaxes into a fundamental state by emission of particles, photons or change of internal energy, for example, into heat. This relaxation can be detected, for example, by optical or mass spectroscopy.
- Excitation of the initial state by one probe pulse into an intermediate state. This new state is probed by an additional probe pulse. For example, the first excitation by the probe beam increases the cross-section of absorption and a secondary process induced by a second probe pulse, like second harmonic generation (SHG) detects the state of matter.

The generation and detection of a certain state requires a knowledge of the complete radiation parameters influencing these states. Thus, for example, the knowledge of the chirp of the radiation is necessary for time-resolved spectroscopy in order to correlate the spectral components of the radiation with the time.

In addition to the preparation of the pump and probe experiment by choice of one parameter of the laser radiation, preparation can also be done by temporally shaping the probe radiation. Temporally shaped radiation interacts selectively with matter enabling measurement of an excited state. The selectivity is given by the adapted electrical field of the radiation to the chemical bonding properties of the investigated matter. Like the key-lock, temporally shaped laser radiation exhibits defined reactions. With definite chemical or physical reactions, like the bond-breaking of molecules, ablation by evaporation without melting could be initiated.

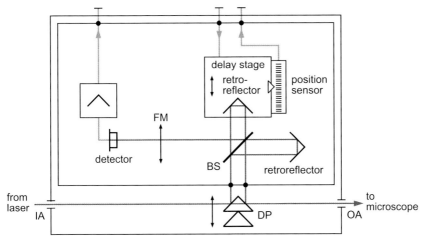

Fig. 4.3 Set-up of an autocorrelator for the measurement of unfocused and focused ultra-fast laser radiation.

4.2.4
Measurement of Radiation Properties

4.2.4.1 Pulse Duration

Ultra-fast laser radiation exhibits extraordinary properties in addition to the spectral and the spatial intensity distribution, which have to be measured in order to get a definite assignation to the pump and probe experiment: the temporal distribution and the phase of the probe radiation. These two process parameters have to be pre-set supplementary in order to obtain the desired state of matter.

The pulse duration for ultra-fast laser radiation cannot be measured directly, because the temporal distribution of this ultra-fast radiation cannot be detected directly (Table 4.1). A correlator uses nonlinear mixing as a combination of filters to generate a signal that can be measured by a slow detector (Figure 4.3). By autocorrelating the pulse duration (FWHM), FROG and GRENOUILLE can be detected (see the next section). Asymmetries of the temporal shape cannot be detected by this method because autocorrelation measurements result in symmetrical temporal profiles of the pulse duration.

Using a cross-correlator the temporal distribution of the radiation is detected with a dynamic range of 10^{10}. This technique enables the detection of asymmetries in the pulse shape and also for measurement of artifacts. Pedestals in the ± 1 ps range, ghost pulses in the range up to 300 ps as pre- and post-pulses, and amplified spontaneous emission are in the nanosecond regime and can be detected by third-order cross-correlation measurements[27].

The pulse duration in the focus is detetcable using an autocorrelator and aligning the focused laser radiation in autocollimation, or by implementing the objective

---

27) http://www.amplitude-technologies.com/
     File/SEQUOIA.pdf

Table 4.1 Commercial autocorrelators (AC) and cross-correlators (CC) for pulse duration, spectrum and phase measurement.

	Manufacturer	Model	Range	Resolution
			ps	%
	APE	Mini	0.01–15	5
	APE	Carpe	0.15–15	5
	APE	PulseCheck	0.15–250	2
	APE	FROG	0.02–1	5
	Coherent	Single-Shot Autocorrelator fs	0.25–0.5	5
AC	Coherent	Single-Shot Autocorrelator ps	0.5–20	5
	Newport	PulseScout Autocorrelator	0.05–3.5	5
	Newport	Long Scan Autocorrelator	0.05–160	0.1
	Swampoptics	Grenouille	0.01–0.1	5
	Swampoptics	Grenouille	0.5–5	5
	Thales Lasers	Taiga Single-Shot AC	0.03–1	5
CC	APE	SPIDER	0.04–0.15	5
	Amplitude	Sequoia	0.05–0.25	10^{-8}

into the Michelson interferometer of an autocorrelator, substituting the focusing mirror (FM) and the internal detector with the microscope objective and an external detector. This autocorrelation is compared to the measurement without objective (Figure 4.3). Autocorrelators detect pulse durations in the range of $5\,\text{fs} < \tau_p < 100\,\text{ps}$.

4.2.4.2 Spectral Phase

The spectral phase is measured indirectly using cross-correlation (Eq. (4.5)) of the fundamental and the SH of the laser radiation. The method used is called Spectral Phase Interferometry for Direct Electric Field Reconstruction[28] (SPIDER)[29]. It is an interferometric measurement technique in the frequency domain based on spectral shearing interferometry. Spectral shearing interferometry is related to intensity autocorrelation with the difference that instead of gating a pulse with a delayed equal pulse, a pulse is interfered with a frequency-shifted or spectrally sheared copy of itself. It provides an opportunity for real-time measurement on spectral and time resolved intensity and on the phase of ultra-fast laser radiation.

Frequency-Resolved Optical Gating (FROG) is a derivative of autocorrelation, which can additionally determine the phase of the radiation. In the most com-

28) http://www.ape-berlin.de/gb/products/carpe.
   html
29) http://www.ape-berlin.de/gb/products/
   spider.html

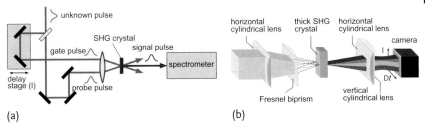

Fig. 4.4 Scheme of a FROG (a) and of a GRENOUILLE (b) [131].

mon configuration, FROG is used as a background-free autocorrelator followed by a spectrometer (Figure 4.4a).

Grating-Eliminated No-nonsense Observation of Ultra-fast Incident Laser Light E-fields (GRENOUILLE) is based on the SHG FROG (Figure 4.4b). GRENOUILLE replaces the beam splitter, delay line, and beam recombination components of the autocorrelator with a prism while the spectrometer and thin SHG crystal combination is replaced with a thick SHG crystal. The effect of these replacements is to eliminate all sensitive alignment parameters while at the same time increasing the signal intensity and reducing the complexity and cost. Like the FROG systems, GRENOUILLE determines the full phase and intensity data of a pulse [131].

4.3
Imaging with Ultra-fast Laser Radiation

The description of some of the fundamentals of optics necessary for pump and probe metrology, like imaging by microscopy, is given in the following sections [62]. The application of pump and probe methods on the micrometer scale requires the knowledge of the theory of image formation in a microscope. Generally, in order to describe the propagation of radiation through objectives, diffraction theory is required. The resolution of an imaging system determines the maximum detectable information of an object. Its dependence on the coherence of the applied radiation is also investigated.

The image formation of an optical system can be calculated for small apertures and are described in Section 4.3.1. The scalar theory used is insufficient to describe the beam propagation in microscope objectives. The complex vectorial description to describe the propagation of electromagnetic field in large Na optical systems is omitted, but results on the resolution power of microscope objectives using non-coherent and coherent radiation are given in Section 4.3.2.

4.3.1
Diffraction Theory and Incoherent Illumination

Every imaging system has a limited aperture which diffracts the incoming light, so the aperture of an imaging system defines the resolving power. For an infinite aper-

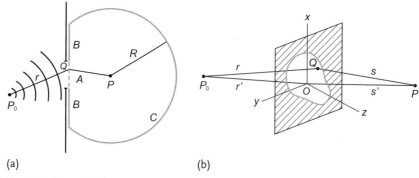

Fig. 4.5 Scheme of diffraction at an aperture (a) and scheme for the derivation of the Fresnel–Kirchhoff diffraction formula (b) [62].

ture $\mathcal{A}$ or an infinitesimally small wavelength λ geometrical optics is valid [132]. The function describing the response of an imaging system to a point source, called a point spread function (PSF), is derived from diffraction theory. The degree of spreading (blurring) of the point object is a measure of the quality of an imaging system. In incoherent imaging systems such as fluorescent microscopes, telescopes or optical microscopes, the image formation is linear and is described by linear system theory. As a result of the linearity, the image of any object in a microscope can be treated as divided into discrete point objects of varying intensity. The image is computed as a superposition of the PSF of each object point. As the PSF is typically determined entirely by the imaging system, the entire image can be described by knowing the optical properties of the imaging system.

In scalar theory the resolving power of an image shaping system – as one possibility – defined as the smallest detectable distance (or angle of observation) of two point sources, is calculated by the Fresnel–Kirchhoff formula derived from the Huygens–Fresnel principle

$$U(P) = \frac{Ae^{ikr_0}}{r_0} \iint_W \frac{e^{iks}}{s} K(\chi)dS, \tag{4.14}$$

where s denotes the distance from the point of observation to a point P on the boundary W, $K(\chi)$ is the inclination factor (in other words, angle of diffraction) and A/r_0 is the amplitude of the incident radiation at P (Figure 4.5a) [62]

$$U(P) = -\frac{Ai}{2\lambda} \iint_A \frac{e^{ik(r+s)}}{rs} \left[\cos(n, r) - \cos(n, s) \right] dS. \tag{4.15}$$

The Fresnel–Kirchhoff formula predicts, when the boundary conditions are given, the amplitude of an electromagnetic wave disturbed by an aperture $\mathcal{A}$ at a point P, arising from the superposition of secondary waves that proceed from a surface situated between this point and the light source (Figure 4.5a). Variables r and s are the distances between the source P_0 and one point in the aperture Q (called entrance pupil), and between Q and the point under investigation, P. $\cos(n, r)$ and $\cos(n, s)$ specify the cosine of the angles spanned by the normal vector $\vec{n}$ with the vectors $\vec{r}$ and $\vec{s}$ (Figure 4.5b).

For optical systems with small numerical aperture the distance between the points P_0 and P, Eq. (4.15), are large compared to the dimension of the aperture. This results in a factor $\cos(n, r) - \cos(n, s)$ and distances r and s that do not vary appreciably over the aperture, and can be summarized by $2\cos\delta$ and the distances r' and s' defined as the distances between P_0, P and the origin of the aperture $\mathcal{A}$ (Figure 4.5b). The Fresnel–Kirchhoff formula (4.15) becomes

$$U(P) \propto -\frac{Ai\cos\delta}{\lambda} \frac{1}{r's'} \iint_{\mathcal{A}} e^{ik(r+s)} dS. \tag{4.16}$$

The points P_0, P and Q and the distances r, r', s, s' are described using a coordinate system in the aperture $\mathcal{A}$ with ξ and η giving the position of Q

$$r^2 = (x_0 - \xi)^2 + (y_0 - \eta)^2 + z_0^2 \quad r'^2 = x_0^2 + y_0^2 + z_0^2 \tag{4.17}$$

$$s^2 = (x - \xi)^2 + (y - \eta)^2 + z^2 \quad s'^2 = x^2 + y^2 + z^2 \tag{4.18}$$

resulting in

$$r^2 = r'^2 - 2(x_0\xi + y_0\eta) + \xi^2 + \eta^2 \tag{4.19}$$

$$s^2 = s'^2 - 2(x\xi + y\eta) + \xi^2 + \eta^2. \tag{4.20}$$

Since the aperture is assumed to be small compared to r' and s', the distances r and s can be expanded [62] to

$$r \propto r' - \frac{x_0\xi + y_0\eta}{r'} + \frac{\xi^2 + \eta^2}{2r'} - \frac{(x_0\xi + y_0\eta)^2}{2r'^3} - \cdots \tag{4.21}$$

$$s \propto s' - \frac{x\xi + y\eta}{s'} + \frac{\xi^2 + \eta^2}{2s'} - \frac{(x\xi + y\eta)^2}{2s'^3} - \cdots \tag{4.22}$$

resulting in an approximation of the Huygens–Fresnel Eq. (4.15)

$$U(P) = \underbrace{-\frac{i\cos\delta}{\lambda} \frac{Ae^{ik(r'+s')}}{r's'}}_{C} \cdot \iint_{\mathcal{A}} e^{ikf(\xi,\eta)} d\xi d\eta, \tag{4.23}$$

with the approximation

$$f(\xi, \eta) = -\frac{x_0\xi + y_0\eta}{r'} - \frac{x\xi + y\eta}{s'} + \frac{\xi^2 + \eta^2}{2r'} + \frac{\xi^2 + \eta^2}{2s'}$$
$$- \frac{(x_0\xi + y_0\eta)^2}{2r'^3} - \frac{(x\xi + y\eta)^2}{2s'^3} + \cdots. \tag{4.24}$$

Defining the direction cosines

$$l_0 = -\frac{x_0}{r'}, \quad l = \frac{x}{s'}$$

$$m_0 = -\frac{y_0}{r'}, \quad m = \frac{y}{s'} \tag{4.25}$$

the Eq. (4.24) becomes

$$f(\xi, \eta) = (l_0 - l)\xi + (m_0 - m)\eta \tag{4.26}$$

$$+ \frac{1}{2}\left[\left(\frac{1}{r'} + \frac{1}{s'}\right)(\xi^2 + \eta^2) - \frac{(l_0\xi + m_0\eta)^2}{r'} - \frac{(l\xi + m\eta)^2}{s'}\right] + \cdots$$

and using $p = (l_0 - l)$ and $q = (m_0 - m)$, considering only the linear part of $f(\xi, \eta)$, one finally obtains the far-field diffraction from Eq. (4.16), by the Fraunhofer diffraction formula

$$U(P) = C \iint_A e^{-ik(p\xi + q\eta)} d\xi d\eta . \tag{4.27}$$

C is defined in terms of quantities depending on the position of the source and the point of observation summarized in Eq. (4.23), ξ and η are coordinates of a point in the aperture. Considering also the quadratic terms in Eq. (4.24) the Huygens–Fresnel diffraction is described for larger NA.

The Fraunhofer diffraction at a circular aperture with radius a results in [62]

$$U(P) = CD\frac{2J_1(kaw)}{kaw}, \tag{4.28}$$

with the definition of polar coordinates

$$p = w\cos\psi, \quad q = w\sin\psi \tag{4.29}$$

and $w = \sqrt{p^2 + q^2}$, $D = \pi a^2$ and J_1 the Bessel function represented by

$$J_n(x) = \frac{i^{-n}}{2\pi} \int_0^\infty e^{ix\cos a} e^{ina} da . \tag{4.30}$$

The intensity distribution itself of an image defined as

$$I = |U(P)|^2 \tag{4.31}$$

is also called point spread function (PSF) described at the beginning of the section. The first minimum in the focal plane $\mathcal{F}$ is given by Abbe's equation

$$w = 0.61\frac{\lambda}{a} \tag{4.32}$$

defining the resolving power of a telescope[30].

30) Because of the large distance of, for example, stellar objects, w gives the minimal detectable angular separation of two stellar objects.

4.3.2
Image Formation by Microscopy and Coherent Illumination

In the theory of resolving power (Section 4.3.1) the formation of an image by an objective has been analyzed by looking at the Fraunhofer diffraction taking into account incoherent light sources. For many self-luminous objects the assumption is valid that the detected intensity at any point in the image plane is equal to the sum of the intensities of the diffraction pattern of each object.

Using a microscope as an observing method objects are usually non-luminous and are transilluminated by an auxiliary system. The auxiliary system, being an illuminating system, for example, a condenser, generates by its aperture diffraction patterns from each element of the source in the image plane of the microscope. The diffraction patterns close to the other one, partly overlap and, being from neighboring points of the source, partly correlate. In general, it is impossible to obtain, by means of a single observation with optical microscopy, a faithful enlarged picture revealing all small-scale structural variations of an object [62]. Using an adapted illumination like Köhler illumination, the resolution power of the imaging system is given for a non-coherent illumination, whereas using coherent illumination the diffraction patterns reduce the resolution power.

Assuming a point Q of a self-luminous object in the object plane Π with the distance Y to the point on the optical axis P and the images Q' and P', the aperture angle of the beam pencil in the image space is

$$\theta' = \frac{a'}{D'},$$
(4.33)

with a' the diameter of the exit pupil at the back focal plane $\mathcal{F}'$ and D' the distance between the back focal plane $\mathcal{F}'$ and the image plane Π' (Figure 4.6). For small focal lengths f the aperture angle w defined by the points Q' and P' with the center of the diffracting aperture describes in good approximation the distance

$$Y' = wD'.$$
(4.34)

For a circular aperture, Eq. (4.32) is valid

$$Y' = 0.61\lambda'\frac{D'}{a'} = 0.61\frac{\lambda'}{\theta'} = 0.61\frac{\lambda_0}{n'\theta'}.$$
(4.35)

Microscopes generate images of objects which are not only on an axial point but are also located off axis. To fulfill this the Abbe sine condition, Eq. (B10), for unity magnification

$$nY\sin\theta = -n'Y'\sin\theta'$$
(4.36)

has to be taken into account. For microscopes θ' is small, thus $\sin\theta' \approx \theta'$, and the minimal distance of an object also defining the resolving power is given by

$$|Y| \approx 0.61\frac{\lambda_0}{n\sin\theta} = 0.61\frac{\lambda_0}{NA}$$
(4.37)

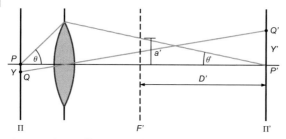

Fig. 4.6 Scheme of beam pencils and resolving power [62].

with the numerical aperture $NA = n \sin \theta$ of the microscope objective. This resolving power is defined for incoherent illumination, for example, by self-luminescent objects. For pump and probe metrology this definition of resolution power is valid, for example, to detect laser-induced fluorescence or laser-induced plasma emission.

Pump and probe metrology of non-luminescent objects is the main field for the detection of density states of gases, plasmas and solid state objects, like the change of the state of matter. These objects are transilluminated by the probe radiation. As a matter of fact, this radiation is coherent laser radiation. The temporal coherence $\Delta t = c/\Delta l$ is dictated by the pulse duration of the probe radiation and the spatial coherence is maximal and unity.

The determination of the resolution with coherent illumination can be done by Abbe's theory. Coherent illumination is achieved by reducing the dimension of an incoherent illumination source[31] or using laser radiation. Not only does the aperture of the objective diffract light, but the object also acts as a diffracting grating. Mathematically speaking, the Fraunhofer diffraction of the object is calculated by Eq. (4.27) for the focal plane $\mathcal{F}'$ and then the Fraunhofer diffraction of this is calculated for the aperture of the objective (for more details see [62]). Finally, for the limit of resolution with coherent illumination one obtains

$$|Y| = 0.82 \frac{\lambda_0}{n \sin \theta} .$$ (4.38)

Up to now only imaging of point sources has been discussed and will be adopted for extended objects. As shown in [62] the resolution power of a microscope for incoherent illumination of an extended object is described by Eq. (4.35) approximately. For coherent illumination of an extended object the resolution becomes

$$|Y| > \frac{\lambda_0}{n \sin \theta} ,$$ (4.39)

due to the interference of neighbor points of the object generating diffraction points that are twice the value for incoherent illumination. This is the main

---

31) Spatial coherence of the illumination is achieved by reducing the aperture diameter of the condenser

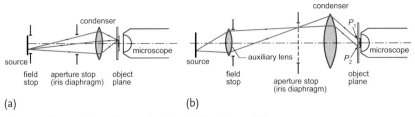

Fig. 4.7 Schemes of critical (a) and Köhler (b) illumination [62].

handicap of using coherent laser radiation as probe radiation. Alternatives like low-coherence high-brightness LED, or an ultra-fast flash lamp can be adopted, but these sources usually suffer from a too large pulse duration and a too small intensity, which is, however, necessary for the detector. Also, because of the partially coherent properties of this radiation, a closer look at the mutual coherence is needed (Section 4.1.3) [62].

In order to examine tie-resolved small non-luminous objects pump and probe techniques are adopted. The illumination method is achieved by transillumination with a condenser. Two methods are described in literature for incoherent radiation: critical and Köhler illumination (Figure 4.7). Coherent laser radiation can only adopt the critical illumination.

The resolving power of an imaging system depends only on the degree of coherence of the radiation and on the resolution power of the objective. Hence, the aberrations of the condenser do not influence on the resolution power of a microscope [62]. Also, it can be demonstrated that the resolution power can be increased when the radiation is incoherent and the ratio of the numerical aperture of the condenser to the numerical aperture of the objective is about 1.2. For coherent illumination no improvement in resolution power can be achieved.

4.4
Temporal Delaying

The vast majority of applications involving ultra-fast lasers for pump and probe experiments require an adjustable time delay for the optical pulses in order to detect the fast processes as time-resolved. Temporal delays are an indispensable part of ultrashort spectroscopy, biological and medical imaging, fast photometry and optical sampling, THz generation, detection, and imaging, and optical time-domain reflectometry [133].

Laser-induced processes can be investigated by pump and probe experiments and need, therefore, a variable delay setting of the probe to the pump pulse. The delay range depends on the process investigated and varies between 1 ps for electric processes (the process duration of chemical reactions) and 1 ms for phonon processes (the process duration of heating, melting). Also dependent of the process duration is the temporal resolution. In general the adopted pulse duration of the

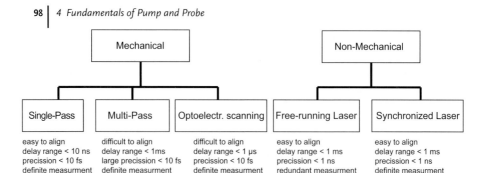

Fig. 4.8 Mechanical and non-mechanical delay set-ups with properties.

probe radiation should be below the process duration of about one magnitude. The parameters of the probe radiation should not vary with changing delay.

A temporal delay can be achieved by a mechanical (Section 4.4.1) or a non-mechanical delay (Section 4.4.2). In the following sections exemplary set-ups are described for both delay classes (Figure 4.8). Each set-up is characterized by the following properties:

- alignments effort: easy or difficult to adjust,
- delay range: measurement range of the pump and probe experiment,
- precision of the delay line: resolution of the delay line, and
- productivity of the measurement: a deliberate delay or a redundant delay is measured.

4.4.1
Mechanical Delay

4.4.1.1 Single-Pass Delay Stage
The conventional method for delaying a pulse is to separate the laser radiation with a beam splitter and direct it to a movable mirror and back. The distance between beam splitter and mirror is D. This results in a delay

$$\Delta t_{\text{delay}} = \frac{2D}{c},$$ (4.40)

with c the speed of light. Using stepping motors, the position of the mirror is changed and a spatial resolution of 0.1 µm is easily achievable today. This results from using Eq. (4.40) in a temporal resolution of the delay line of $\approx 1\,\text{fs}$[32]. For generation of a delay of about 10 ns a delay line length of about 1.5 m is necessary. All mechanical delay lines are highly sensitive in the adjustment. In order to reduce the sensibility, corner cubes are used for single-pass delay lines. For larger delay lines it has to be considered that the confocal parameter of the laser radiation, defined as two times the Rayleigh length z_R, Eq. (2.10), is about 10 times larger than the delay

---

32) not to be confused with the temporal
   resolution given by the pulse duration of the
   probe radiation

line length. For a delay line length of about 3 m a confocal parameter of 30 m is needed, resulting in beam diameter of about 10 mm for 1 μm laser radiation. The change of the focal position z at the microscope objective as function of the delay due to the changed position of the beam waist parameters can be neglected.

4.4.1.2 Multi-Pass Delay Stage

Larger delays into the μs-regime are feasible using a self-reproducing Herriott resonator as a multi-pass delay line. The optical quality of the mirror surfaces has to be very good: surface deformations $\ll \lambda/10$ over the beam and a reflectivity very close to unity are necessary. The coatings of these mirrors have to be chosen appropriately for ultra-fast laser radiation in order to minimize GVD and TOD.

The Herriott cell allows a number of beam geometries useful for various applications such as interferometry, spectroscopy and high precision photometry. The design of a Herriott cell for pump and probe applications is described in [134, 135]. The key advantages of the use of multi-pass delay stage for femtosecond pump-probe experiments include:

- keeping pulse duration constant and resulting in short illumination times due to nearly zero-chirp concave mirrors, and
- preserving the beam parameter, as a result of periodic focusing inside the cavity, using appropriate design conditions.

In its simplest form, a Herriott cell-based delay stage consists of a stable two-mirror resonator and an optical element for injecting and extracting light beams, for example, by use of additional flat or curved mirrors, holes or slits in one or both of the mirrors (Figure 4.9). When the Herriott cell parameters are properly adjusted, the incident beam undergoes multiple reflections before leaving the cell with the same beam parameters as the incident radiation. A multi-pass delay stage conceived for up to 300 reflections per mirror is described. Spherical mirrors with focal length $f = 1$ m, diameter $d = 0.15$ m and a 4.5×25 mm^2 radial slit in one of the mirrors have been used. Optical coatings provide high reflectivity ($R > 99.9\%$) and neglecting group delay dispersion (GDD) in the wavelength regime $\lambda = 780 - 820$ nm. The

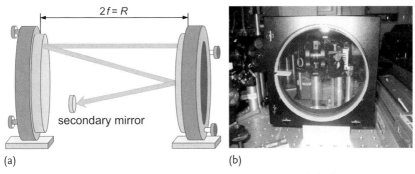

(a)   (b)

Fig. 4.9 Schematic setup of a Herriott delay line (a) and one mirror with slit (b).

distance between two mirrors can be changed in the interval $L \in [f, 2f]$. In order to derive the general condition required to preserve the beam propagation parameter of the Gaussian beam, for an arbitrary Herriott cell, an initially injected beam $\vec{r}_0$ and its transfer matrix M_T at a spherical mirror surface are considered

$$\vec{r}_0 = \begin{pmatrix} r_0 \\ r'_0 \end{pmatrix}, \quad M_T = \begin{pmatrix} 1 & 0 \\ -2/R & 1 \end{pmatrix} \quad M = \begin{pmatrix} 1 & L \\ 0 & 1 \end{pmatrix} M_T. \tag{4.41}$$

r_0 represents the initial beam displacement from the optical axis, r'_0 is the initial entrance angle of the beam and R is the radius of curvature of the mirrors. After n reflections, the beam vector $\vec{r}_n$ becomes

$$\vec{r}_n = \begin{pmatrix} r_n \\ r'_n \end{pmatrix} = \underbrace{M_1 \cdot M_2 \cdots M_n}_{M^n} \cdot \begin{pmatrix} r_0 \\ r'_0 \end{pmatrix}. \tag{4.42}$$

The Heriott cell can be described in the Jordan–Bloch formulation by

$$M^n = \begin{pmatrix} L & 0 \\ L/R & \sin\phi \end{pmatrix} \begin{pmatrix} \cos(n\phi) & \sin(n\phi) \\ -\sin(n\phi) & \cos(n\phi) \end{pmatrix} \begin{pmatrix} \frac{1}{L} & 0 \\ -\frac{1}{R\sin\phi} & \frac{1}{\sin\phi} \end{pmatrix} \tag{4.43}$$

Because of typically small entrance angles and very large radii of curvature of the mirrors, astigmatism resulting from the tilted beam can be neglected [136]. The transverse displacements x_n and y_n of the beam and the angular position θ of the spots around the ellipse formed after n successive reflections can be given

$$x_n = A\sin(n\theta + \alpha), \quad y_n = B\sin(n\theta + \beta), \quad \cos\theta = 1 - \frac{L}{R}. \tag{4.44}$$

Thus, if

$$A = B = R_H = \sqrt{x_0^2 \left(1 + \frac{L}{R}\right) + LR x_0'^2 - 2Lx_0 x_0'}, \tag{4.45}$$

a circle of radius R_H will be described by the reflecting beam. In order to preserve the beam parameters of the Gaussian radiation, for the specific case of $y'_0 = 0 \rightarrow \beta = 0$, following start conditions the initial beam displacement and entrance angle can be given

$$x_0 = R_H, \quad x'_0 = -\frac{R_H}{R} \quad \text{and} \quad \alpha = -90° \rightarrow \frac{x'_0}{x_0} = \frac{1}{R}. \tag{4.46}$$

In this specific case, in order to satisfy the periodic "re-entrance condition" [137], the following boundary conditions have to be noted

$$N \cdot \theta = 2\pi j, \quad \text{and} \quad L = 2f \left(1 - \cos\frac{2\pi j}{N}\right). \tag{4.47}$$

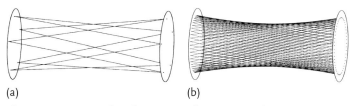

(a)   (b)

Fig. 4.10 Beam tracing for a the Herriott cell (ROC = 2 m) for
N = 10, L = 1382.0 mm (a), and N = 100, L = 1749.3 mm (b).

N is the number of beam passes inside the Herriott cell, and j is an integer representing the round-trip number of the spots around the circle. Using Eqs. (4.44)–(4.47) calculations for the mirror separation length can be performed, according to required numbers of beam passes and beam tracing using a numerical approach results in a maximal delay of $t_{delay} = 1.8\,\mu s$ (Figure 4.10).

About 300 passes within a Herriott delay line have been realized with an overall transmissivity of 50% resulting in a delay time $t_{delay} = 1.805\,\mu s$ [134, 135].

4.4.1.3 Mechanical Optoelectronic Scanning

In order to increase the productivity and the sampling rate, scanning of the delay is introduced. The laser radiation is reflected at stepped mirrors using acousto-optic deflectors as a dispersive element [138]. Mechanical scanning is achieved by shakers, rotating mirrors and linear translators using stepper motors and galvanometers.

The temporal resolution of a mechanical delay is about 10 fs (Section 4.4.1.2). The achieved scanning rate is low due to mechanical limitations by resonance effects and the inertial mass of the rotating mirrors. Scanning ranges using galvanometer between 50 fs and 300 ps at 30 Hz scanning frequency have been achieved [139]. The advantage of rapid scanning has been pointed out in [139], demonstrating that fast scanning moves the data acquisition frequency range out of the baseband noise of the laser source, which has been demonstrated up to 400 Hz scanning frequency, indicating the mechanical limit.

4.4.2
Non-mechanical Delay

Different approaches for non-mechanical delaying by scanning have been realized. The mechanical systems using scanning are the most used historically, but also the slowest concepts. A delay is achieved either by:

- deflecting the laser radiation into a delay line using an electro-optical or acousto-optical modulator, or
- a second illumination source as a probe beam.

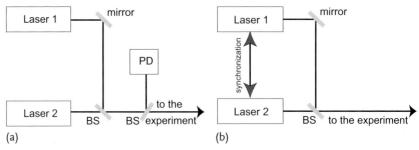

Fig. 4.11 Set-up for free-running (a) and synchronized laser (b).

Deflecting radiation is difficult to align and will not be described in this book. Non-mechanical delay by using a second illumination is achieved by coupling two or more laser sources, which are free-running or synchronized.

4.4.2.1 Free-Running Lasers

Using free-running lasers, defined as a non-temporally synchronized laser with an illumination source, enables very fast delaying and scanning. Using a flashlight or LED as a probe beam is the conventional method in pump and probe metrology, having the drawback of too large pulse durations above 100 ns in the case of large luminescence. Some flash lamps and LEDs generate radiation with a pulse duration $t \approx 10$ ns, but with too small luminescence. Some excimer lamps achieve sub 100 ns pulse durations, but the resulting fluence is also very small. So the proper choice is the use of an ultra-fast laser source.

A free-running set-up is achieved by aligning the laser sources coaxially using dichroitic beam-splitters: laser one delivers the pump beam and laser two the probe beam (Figure 4.11a). A temporal causality between the two sources is achieved by detecting the radiation of both lasers during the experiment by a photo diode.

Pulse time delaying can be produced by two free-running laser systems having different repetition rates. This set-up produces a time delay sweeping effect, with the sweep repetition frequency at the offset frequency between the repetition rate of the two laser systems. Free-running laser systems can achieve a large scan frequency and slew rate. For example, by using two laser systems working at $f_p \approx 80$ MHz and having slightly different cavity lengths, a scanning range of 13 ns can be achieved. Accurate timing calibration is achieved by cross-correlating the laser radiation of the two sources (Section 4.1.3).

When the two laser sources have different repetition rates of orders of magnitude the approach by free-running systems is still feasible. For example, the combination of an ultra-fast fiber laser system at 1 MHz has been coupled to a Ti:sapphire laser source working at 0.5 Hz for the detection of the optical phase changes induced in glass during laser welding (Section 6.4) [140].

A region of interest is desired, for example at 80 MHz only the range between $0 < t < 10$ ps has to be investigated. The main drawback of this method is the redundant data generated. Statistically all temporal points are scanned, but often only the data in the regime between 10 ps and 12 ns given by the experiment is not within the range of interest. By increasing the repetition rate into the MHz-regime, the regime of not-useful data can be reduced using the sweeping regime, but the approach is not flexible for small delays.

4.4.2.2 Synchronized Lasers

Often the lasers are connected to an external master clock or are synchronized by an internal clock. Synchronized lasers are achieved by aligning the laser sources coaxially using dichroitic beam-splitters: laser one delivers the pump beam and laser two the probe beam (Figure 4.11b). Synchronization is achieved by:

- passive optical methods, or
- electronic synchronization.

The passive optical method enables the largest accuracy – at least as small as the used pulse duration – through interaction of the laser radiation of both sources by an optical effect, for example, cross-phase modulation.

Electronic stabilization using RF phase detection represents a flexible tool in terms of adjustment of time delays. The time accuracy given by the jitter of the electronics is on the order of a few ps. A hybrid optoelectronic method, called pulse phase-locked loop, enables timing stabilizations < 100 fs, being also "non-flexible" like passive optical methods.

A promising approach for electronic synchronization is the asynchronous optical sampling (ASOP[33]) by scanning with two laser systems [133]. This set-up, like the free-running laser scanning system, consists of two lasers, a master and a slave, that have nearly identical cavity lengths and repetition rates. The cavity length of the slave laser is controlled by a PZT[34]-mounted cavity mirror. Unlike a free-running system, the pulses of the master and the slave laser are not scanned through each other in the complete scan regime. The two lasers are synchronized with a phase-locked loop circuit. While repetition rate of the master laser is fixed at a constant frequency ω_1, the repetition rate of the slave ω_2 dithers around ω_1 by periodically varying the cavity length of the slave laser at frequencies between 30 Hz and 1 kHz.

Assuming that the PZT is evolved with a square wave function Sq, with $-1 \leq Sq(x) \leq 1$, at the frequency f_s then the cavity length mismatch is given by

$$\Delta L(t) = \Delta L_0 \cdot Sq(f_s t) , \tag{4.48}$$

where L_0 is the amplitude of the square wave displacement. Consequently one gets the time-varying delay

$$T_D(t) = \frac{1}{L} \int_0^t \Delta L(t') dt' . \tag{4.49}$$

33) http://www.menlosystems.com
34) PZT: Lead zirconate titanate (Pb[Zr$_x$Ti$_{1-x}$]O$_3$, $0 < x < 1$) a perovskite is featuring the piezoelectric effect. Being piezoelectric, it develops a voltage difference across two of its faces when compressed, or physically changes shape when an external electric field is applied.

The time-varying delay is a linear sawtooth function, with an temporal regime

$$\Delta T_{max} = \frac{\Delta L_0}{2L} \left(\frac{1}{f_s} \right) .$$ (4.50)

The scan rate, or sweeping rate, is given by

$$f \equiv \frac{\partial T_D(t)}{\partial t} = \frac{\Delta L}{L} ,$$ (4.51)

and the scan velocity

$$v(t) = \frac{c\Delta L(t)}{L} .$$ (4.52)

For example, working with laser radiation with a 1 GHz repetition rate, a scan velocity of 3 km/s for a cavity length displacement of 1.5 μm is achieved. The scan range is also dependent on the repetition rate of the laser source and the scan frequency. For a pair of lasers emitting laser radiation at 5 MHz a scan range of 10 ns at a scan frequency of 25 Hz has been achieved [133].

5
Examples for Ultra-fast Detection Methods

The laser radiation is divided in one pump beam and at least one probe beam by a beam splitter (Figure 5.1). The one probe beam can, for example, excite the substrate, and the second one detects the change.

Using the pump and probe technique for the process visualization of laser-induced processes, two different approaches are possible (Figure 5.2):

- non-imaging detection (Section 5.1), and
- imaging detection (Section 5.2).

Depending on the detector and the optics used for measuring the probe beam, a two-dimensional image of the investigated area or just the an spatially averaged information can be visualized.

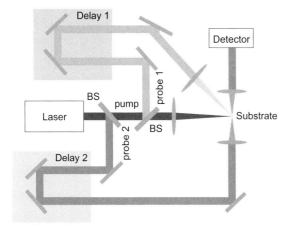

Fig. 5.1 Principle setup of a pump and probe experiment with pump, probe 1, and probe 2 beam.

Ultra-fast Material Metrology. Alexander Horn
Copyright © 2009 WILEY-VCH Verlag GmbH & Co. KGaA, Weinheim
ISBN: 978-3-527-40887-0

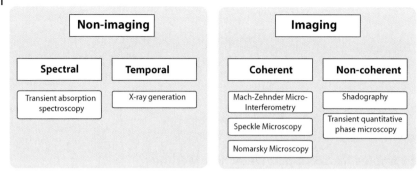

Fig. 5.2 Non-imaging and imaging detection.

5.1
Non-imaging Detection

Two examples are given for non-imaging detection methods. The first using spectral properties of the probe radiation, transient absorption spectroscopy, is a "classical", un-correlated, ultra-fast technique applied in many fields of research. The central feature is the use of selective radiation to detect spectral properties of a laser-induced process. A focus will be given to the ultra-fast white-light continuum (Section 5.1.1).

The second detection method modulates the temporal profile of the probe radiation and generates ultra-fast X-rays. By using double pulses and varying the process parameters, like pulse energy, delay time, and pulse energy ratio, X-ray (Si-K_α radiation, explained in more detail below) is generated with the maximization of efficiency using double pulses. A correlated technique is described to maximize the generation of this radiation (Section 5.1.2).

5.1.1
Transient Absorption Spectroscopy (TAS)

5.1.1.1 Principle and Setup
Transient absorption spectroscopy is a pump and probe technique detecting the absorptivity of a sample using radiation at specific wavelengths generated by non-linear processes of the probe radiation. Selectivity of the probe radiation absorption is given by the resonance bands exhibited by the atoms, molecules or ions in the interaction-zone of the investigating sample. In combination with the lock-in technique a large signal-to-noise ratio can be obtained (Addendum A).

Using as probe radiation broad spectral radiation, called white-light continuum, the induced processes can be investigated in the spectral range from about 300 nm to 2000 nm. In this way ultra-fast white-light continuum is used, generated by self-phase modulation of the ultra-fast laser radiation in a dielectric material (Section 3.1.2.3). As dielectrics sapphire is often adopted, because of its large damage threshold fluence. It enables a white-light continuum with the smallest wavelength

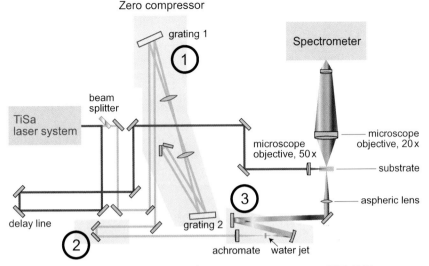

Fig. 5.3 Scheme for an ultra-fast transient absorption spectroscopy setup (TAS) [141].

in the range of 400 nm. The conversion efficiency, being strongly dependent on the pulse duration, is less than 1% for ultra-fast laser radiation with a pulse duration of 80 fs. Larger efficiencies can be achieved using water or CS_2 in a jet with a smallest wavelength in the range of 320 nm. Photonic crystal fibers enable the generation of a white-light continuum with large efficiency, but with the disadvantage of limited intensities. The photonic crystal fiber is a polarization-maintaining supercontinuum device used with femtosecond laser radiation at λ = 800 nm wavelength. It enables the generation of a white-light continuum with zero dispersion at 750 nm wavelength. The fiber ends are sealed and mounted in quartz ferrules enabling a larger damage threshold[35].

The specifications of the spectrometer and the sensitivity of the detector depend on the spectral resolution necessary to resolve the process. Using photo multipliers or iCCDs, single-photon counting is achievable. For reversible laser-induced processes, like photo-induced bleaching, the detected signals can be integrated by multiple measurements. It enables by integration the detection of processes with small signal intensities, even using photo diodes or CCDs. A CCD-based commercial turn-key system featuring pump-probe transient absorption spectroscopy designed to work with an amplified Ti:Sapphire femtosecond laser is available today[36].

Transient absorption spectroscopy is achieved by a pump and probe setup using an ultra-fast laser source (Figure 5.3). A beam splitter divides the beam in a pump and a probe fraction. The pump beam is guided to the sample and focused onto it. The probe beam passes an optional zero-compressor and is delayed in a mechanical

35) http://www.crystal-fibre.com/products/
   femtowhite.shtm
36) http://www.ultrafastsystems.com/helios.htm

delay line. The zero-compressor changes the chirp of the probe beam enabling the use of different pulse durations for pump and probe radiation.

Finally, the probe radiation is focused into a water jet generating an ultra-fast white-light continuum (WLC) [85]. Attention has to be paid to producing a laminar flowing and spatially stable water jet by adopting a high-precision nozzle and a stable water pump. The white-light continuum is collimated by a concave silver-coated mirror and guided through the sample and optics to the entrance slit of the spectrometer.

As described in Section 4.2.3 the properties of the probe radiation have to be detected. This will be given for the white-light continuum in the following, firstly by generating and characterizing the probe radiation (pulse duration, spectral distribution, chirp) and secondly by describing the measurement procedure.

5.1.1.2 Characterization of the White-Light Continuum

Generation of WLC The generation of the WLC is achieved by non-linear processes induced in a water jet. The probe radiation is focused by an appropriate Fraunhofer achromate (Bernhard Halle f = 60 mm, F/♯12) into the water jet. Because of self-focusing induced by the Kerr effect, and defocusing, driven by the free electrons, filaments occur in the water jet. The filaments propagate through the water and by four-wave mixing and self-phase modulation the spectral bandwidth of the WLC is increased (Section 3.1.2.3). At small average power only one filament is formed (P = 0.75 mW, Figure 5.4). With increasing average power (for example, P = 2.3 mW) a second filament is formed which interferes with the first one [142].

The distance between the filaments increases due to thermal effects with increasing average power until both are completely separated ($P \approx$ 5 mW, Figure 5.4). The spatial mean intensities of the two filaments oscillates in time.

Further increase of the average power induces a shift of the beam waist relative to the water jet position toward the water surface and spatially stable interference of two (P = 5.1 mW), three (P = 7 mW) and four filaments (P = 10 mW) can be detected (Figure 5.5). Depending on the position of the beam waist relative to the

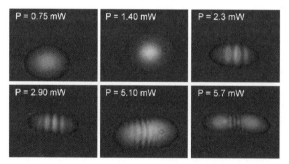

Fig. 5.4 Far-field spatial intensity distribution of WLC as a function of average power (λ = 810 nm, t_p = 70 fs, f_p = 1 kHz, f = 60 mm, F/♯12)[143].

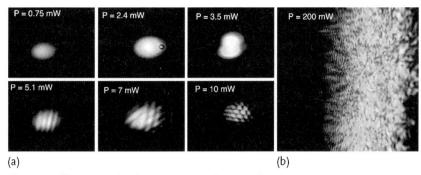

(a)   (b)

Fig. 5.5 Stable intensity distribution of WLC as function of average power and focal position for small average power ($P \leq 10$ mW) (a) and for large average power ($P = 200$ mW) (b) [144].

water jet, the filaments interfere with each other. In the case of the beam waist positioned in front of the water jet, two separated filaments are generated.

The conversion efficiency of the WLC reaches a maximum at the smallest applied pulse duration $t_p = 80$ fs. The largest efficiency varying the circular water jet nozzle diameter ($d_{nozzle} = 0.2, 0.5, 0.8$ and 1 mm) is achieved at the largest diameter. The conversion efficiency reaches its maximum when the ratio of pulse intensity and irradiated volume reaches maximum. For efficient self-phase modulation the pulse intensity has to be $\geq 10^{10}$ W cm^{-2}. The spectral width of the WLC increases. The minimum wavelength of the WLC decreases with increasing intensity towards the UV with an increasing number of filaments (Section 3.1.2.3).

The maximum conversion efficiency is not reached when positioning the beam waist in the center of the water jet, but by positioning 2 mm in front of or behind the water jet (Figure 5.6). A conversion efficiency of $\eta = 1\%$ has been achieved by generation of a large number of filaments ($E_p = 300$ µJ, $t_p = 80$ fs, $\lambda = 810$ nm). These filaments exhibit a diameter of 10–30 µm and exhibit coherence like the fun-

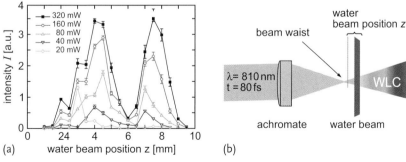

(a)   (b)

Fig. 5.6 Intensity of the WLC as function of the position (a) and scheme of the WLC-generator (b) ($\lambda = 810$ nm, $t_p = 80$ fs) [141].

damental radiation. The spatial intensity distribution in the far-field of the WLC is inhomogeneous and features speckle patterns with different colors (Figure 5.5b). At threshold intensity the water is evaporated and the propagation of the WLC is disturbed and unstable, by diffraction due to shock waves, deflection at bubbles and laser-induced pressure variations of the atmosphere. The resulting speckle patterns change consequently from pulse to pulse.

Pulse duration and chirp of WLC   The pulse duration of ultra-fast laser radiation is measured by an auto-correlator (Section 4.2.4) [145, 146]. The non-linear properties of dielectrics are utilized for the cross-correlation of two laser pulses: irradiating a crystal with two beams interfering in the cross-point under an angle Θ the frequency mixing can be described by the vectorial summation of the wavenumber vectors [147]

$$\vec{k}_{ac} = \vec{k}_1 + \vec{k}_1 \,. \tag{5.1}$$

The angle Θ for the incoming radiation is different from the plane of incidence of the non-linear crystal, and the frequency-converted radiation is selected by an aperture. As described in Eq. (4.5), Section 4.1.3, the correlation between two different laser pulses of radiation possessing different polarization, wavelength or pulse duration is called cross-correlation. The vectorial sum can be written as $\vec{k}_{cc} = \vec{k}_1 + \vec{k}_2$ (Figure 5.7). The resulting wavelength of the cross-correlation can be calculated as

$$\lambda_{cc} = \frac{\lambda_p \lambda_{wlc}}{\lambda_p + \lambda_{wlc}} \,. \tag{5.2}$$

The cross-correlation of the fundamental λ_p = 800 nm with one spectral component of the WLC, for example, with the wavelength λ_{WLC} = 502 nm, results in a cross-correlation radiation with the wavelength λ_{cc} = 310 nm.

The determination of the pulse duration as function of the delay by cross-correlation of the WLC with the fundamental radiation is achieved by detection of the spectral distribution of the resulting cross-correlation signal. A non-linear crystal is used which has a sufficient acceptance angle for all the investigated wavelengths (type I β-Barium-Borate, thickness d = 100 μm). The polarization of pump and probe radiation is set orthogonally to the plane of incidence of the

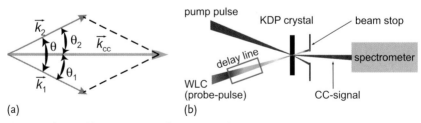

(a)   (b)

Fig. 5.7 Scheme of frequency mixing for cross-correlation (a) and of cross-correlation for WLC with pump radiation (b) [144].

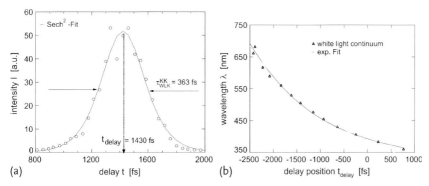

Fig. 5.8 Cross-correlation signal at $\lambda = 310\,nm$ as a function of the delay (a) and wavelength of the WLC as function of the delay position (b) [144, 148].

two beams. The crystal used is optimized for the conversion of radiation at the wavelength $\lambda = 600\,nm$. Consequently, to fulfill the phase-matching condition for cross-correlation of radiation with different wavelengths, the angle of incidences of pump and probe θ_1 and θ_2 has to be changed (Figure 5.7).

The cross-correlation signals from the WLC as a function of the delay are detected (Figure 5.8a). Adapting a sech^2-function at these measured values, the pulse duration of the cross-correlation signal τ_{WLC}^{CC} (FWHM) and the delay position t_{delay} as function of the wavelength are determined. τ_{WLC}^{CC} represents the pulse duration of one spectral component of the WLC at the delay position t_{delay}. The cross-correlation of the WLC at the wavelength $\lambda_{WLC} = 310\,nm$ with the fundamental radiation ($\lambda = 810\,nm$) results in a pulse duration of the cross-correlation signal of $\tau_{WLC}^{CC} = 363\,fs$ at the wavelength $\lambda = 502\,nm$ of the WLC (Figure 5.8a).

The delay position t_{delay} of the WLC changes monotonically as a function of the wavelength (Figure 5.8b), and is a linear function of the frequency (Figure 5.9a). The frequency components of the WLC are chirped with an overall chirp $\Delta t_{chirp} =$

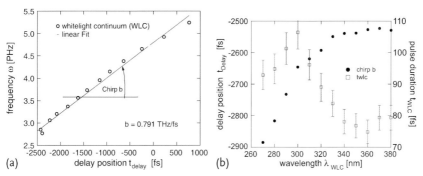

Fig. 5.9 Frequency of WLC as function of delay position (a) and chirp and pulse duration of the WLC as function of wavelength (b) [144, 148].

3500 fs and are induced by a quartz plate used as a high-pass filter and by the water itself. Different frequency components reach the sample under investigation at different times.

The gradient $\mathbf{b}$ of this straight line represents the chirp of the WLC, Eq. (3.13) and results in $b = (0.791 \pm 0.02)$ THz/fs. Because of the linear dependence of the frequency on the delay time

$$\omega(t_{delay}) = a + b \cdot t_{delay} \tag{5.3}$$

a relation

$$\lambda = \frac{2\pi c}{a + b \cdot t_{delay}} \tag{5.4}$$

between wavelength and delay time can be derived. The chirp as function of the wavelength results in

$$\tilde{b}(\lambda) = -\frac{b\lambda^2}{2\pi c} . \tag{5.5}$$

For the wavelength $\lambda = 502$ nm the chirp is calculated as $\tilde{b} = (-0.105 \pm 0.03)$ nm/fs. According to the chirp factor

$$a := -bt_p^2/2 , \tag{5.6}$$

the pulse duration of the WLC t_{WLC}

$$t_{WLC}^{CC} = \tau_{WLC}\sqrt{1 + a^2} \tag{5.7}$$

is chirped to the measured τ_{WLC}^{CC} (Figure 5.8a). The unchirped pulse durations of the WLC t_{WLC} measured at a wavelength are comparable to the pulse duration $t_p = 80$ fs of the fundamental one (Figure 5.9b).

5.1.1.3 Measurement by TAS

The absorption bandwidth of an event can be calculated due to the linearity of the chirp b (Figure 5.9a). The absorption of photons lasting for a very short time $\Delta t = \delta(t)$ is detected as an absorption band in the WLC spectrum (event 1 in Figure 5.10).

For a process with a spectral flat absorption duration $\Delta t \leq t_{chirp}$, the bandwidth of the absorption in the WLC is increased proportionally to the duration of the absorption (event 2 in Figure 5.10). The spectral bandwidth $\Delta\omega_{WLK}$ of each frequency of the WLC is equal to the bandwidth of the generated radiation. The measurement signal

$$A(\Omega) = \int_{-\infty}^{\infty} S(\Omega)R(\Omega, \omega)d\omega \tag{5.8}$$

is a convolution of the frequency distribution $S(\Omega)$ of the WLC with the response function of the WLC-irradiated interaction volume $R(\Omega)$.

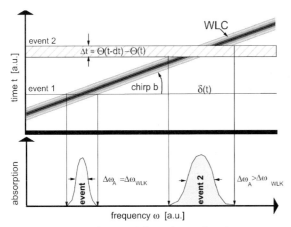

Fig. 5.10 Scheme of temporal dependence of an absorption process from the frequency.

The absorption bandwidth (FWHM)

$$\Delta\omega_A = \frac{1}{\tau_A}\sqrt{8\ln 2(1 + a^2)} \tag{5.9}$$

can be determined assuming a Gaussian spectral distribution of the WLC and a linear chirp $\mathbf{b} = \partial\Phi/\partial t$ [72].

The Eq. (5.9) for the absorption duration τ_A exhibits one real solution. Because the response function of the WLC-interaction region $R(\Omega)$ is generally a Fourier-transformed step-function the following is assumed

$$R(\Omega) = \mathcal{F}\left[\Theta(t + t_{\text{event}}) - \Theta(t)\right] . \tag{5.10}$$

The duration of the event t_{event} results in

$$t_{\text{event}} = \tau_A - \tau_{\text{WLC}}^{CC} . \tag{5.11}$$

5.1.2
Temporal Shaped Pulses

5.1.2.1 Principle and Setup

The generation of X-rays can be adopted for diagnostics, like time-resolved detection of alumina impurities on the surface of silicon wafers [149]. Intense laser radiation can generate X-ray radiation by hot plasma-matter interaction. Structural changes can be observed in a pump and probe setup using these high-brilliance laser-generated X-rays in the picosecond range and the micrometer size [150, 151]. Especially in imaging diagnostics the K_α-emission line enables high-contrast detection.

In the described investigation Si-K_α-radiation is used for the detection of aluminum, because the Si-K_α-radiation exhibits a large cross-section for K-shell ionization of aluminum. The double-pulse laser radiation induced emission of Si-K_α-radiation has been detected using a multi-layer spectrograph. The dependence of

Fig. 5.11 Setup with silicon wafer, photo-diodes and auto-focus.

the emitted Si-K_α-photons on the delay and the energy of the double pulses has been investigated.

For the laser-generated X-ray radiation a MOPA system, based on a diode-pumped Cr:LiSAF oscillator and a diode-pumped Cr:LiSAF and Cr:LiSGAF regenerative amplifier, have been used [152]. The output power of the MOPA system is 100 mW at 1 kHz repetition rate, 100 fs pulse duration, and a beam quality $M^2 = 1.3$. In order to reach intensities of up to 20 PW/cm^2 the laser radiation is focused by a chromatic- and chirp-corrected aspheric lens ($f = 8$ mm) to a calculated beam diameter of about 2 μm (Section 2.2.4). The laser radiation is focused on a silicon wafer with a 4″ diameter and 20 μm thickness (Figure 5.11). The lens itself is protected from ablated material by a moving Mylar foil in front of the lens. In order to irradiate with single pulses, a mechanical shutter is used. A burst of a minimum of 10 pulses irradiates the surface and is achieved by a minimal shutter time 10 ms.

During processing a shift of the wafer surface of < 10 μm occurs. The position of the beam waist (relative focal position) has to be controlled with respect to the Rayleigh length of $z_R = 3.5$ μm. An auto-focus system based on an astigmatic method is used for the compensation of the shift with amplitudes < 10 μm and oscillation frequencies < 100 Hz. All experiments were conducted in a vacuum chamber at a pressure $p < 5 \times 10^{-4}$ mbar. The silicon wafer is moved by x-y-axes at a velocity of 20 mm/s.

Using a delay-line (Figure 5.12a) as a pump and probe setup, two spatially parallel and temporally (separated by up to one nanosecond) laser pulses are focused on the same spot. The two pulses, pre- and main-pulse, exhibit orthogonal polarization[37]. The energy ratio can be deliberately adjusted by using prisms and half-wave plates.

---

37) The pulse energy ratio is given by
$E_{\text{prepulse}} / E_{\text{main-pulse}}$.

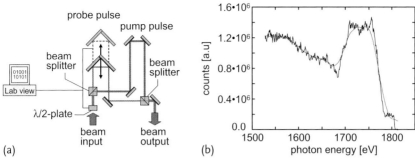

Fig. 5.12 Setup with silicon wafer, photo diodes and auto focus with optics (a), counts of X-ray photons as a function of photon energy for silicon (b) [149].

The measurement of the hot-electron temperature is achieved by two silicon photo diodes detecting X-ray energy within different regimes (Figure 5.11). The photo diodes are positioned radially to the source and the electrical signals S_1 and S_2 are detected.

A multi-layer spectrograph consisting of two multi-layer mirrors and a CCD camera (1340×400 pixels^2) is aligned to detect the X-ray spectra from 1.5 to 1.85 keV with a theoretical spectral resolution of $\lambda/\Delta\lambda = 54$ at $\lambda_{Si-K_\alpha} = 1.739$ keV. Each multi-layer mirror is coated with 100 double-layers consisting of 1.8 nm tungsten and 3 nm silicon and features a maximal reflectivity of about 25% for Si-K_α-radiation at an inclination of 4.4°.

5.1.2.2 Characterization of the Si-K_α-Radiation Source

The generation of X-ray radiation by conversion of IR and visible radiation needs the mediation by electrons: due to resonant absorption of laser radiation in the plasma with a critical electron density $n_{crit}^e(r, t)$, high-energetic electrons, also called hot electrons, can be generated (Section 3.3). The maximal cross-section for Si-K_α-ionization by electrons occurs at 5.5 keV. An ionized atom with a K-shell hole, generated, for example, by high-energetic free electrons, recombines emitting K-shell-photons at an energy of some keV.

The spectrum shows a steadily decaying background of Bremsstrahlung radiation and a characteristic line emission energy by K_α-radiation of silicon at approximately 1.725 keV (Figure 5.12b). From each spectrum the relative number of Si-K_α-photons has been extracted. Because of the high temperature of the plasma the K-shell recombination takes place with partially ionized silicon (Si$^+$–Si^{8+}) with the Si-K_α-line broadened to 90 eV [153].

Delay time The relative number of Si-K_α-photons compared to single-pulse irradiation, can be increased up to 1.5 times by varying the delay between the pre- and main-pulse (Figure 5.13a). A relative maximum has been measured at 36 ps, which can be explained by an optimized coupling of the main-pulse to the plasma.

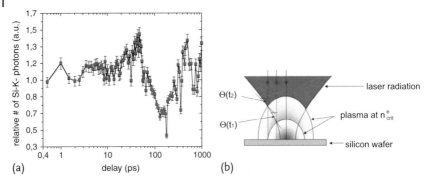

(a)   (b)

Fig. 5.13 Relative number of Si-K_α-photons as a function of delay (pulse energy ratio 5%) (a) and position of plasma surface for the critical electron density n^e_{crit} for two time steps (b) [149].

The critical electron density $n^e_{crit}(r, t)$ of the plasma surface occurs at that delay t_{delay} (Figure 5.13b shown for two delay times).

Pulse energy ratio   The free electrons generated by the ablation of silicon with the radiation of the pre-pulse interact with the radiation of the main-pulse resonantly. Due to the fact that the time where the plasma reaches critical plasma density depends on the pulse energy of the pre-pulse, changing the ratio of the pulse energy means that the delay also has to be varied. By changing the pulse energy ratio between the pre- and main-pulse to about 35%, the number of generated Si-K_α-photons is increased (Figure 5.14a). A maximal number of photons is reached at 35% pulse energy ratio and at about 36 ps delay.

Focal position   The plasma expands and reaches the critical electron density n^e_{crit} about 36 ps after irradiation and at 35% pulse energy ratio (Figure 5.14b). The generation of Si-K_α-radiation by resonant absorption depends strongly on the incli-

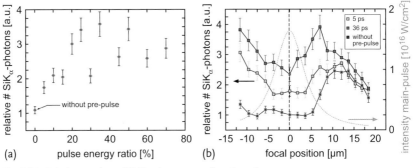

(a)   (b)

Fig. 5.14 Relative number of Si-K_α-photons as a function of pulse energy ratio (delay $t = 38$ ps) (a) and relative number of Si-K_α-photons and intensity of the main-pulse as a function of focal position and delay (pulse energy ratio 36%) (b) [149].

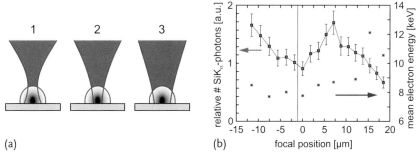

(a)

(b)

Fig. 5.15 Scheme of the focal position 1) focus above 2) on and 3) below the surface (a), relative number of Si-K_α-photons and mean electron energy as a function of focal position (b), delay 36 ps, pulse energy ratio 35% [149].

nation angle of the laser radiation relative to the surface described by the critical plasma density (Section 3.3.1). By varying the z-position of the laser focus relative to the surface the inclination angle, given in this experiment by the divergence angle θ of the focused laser radiation, is changed (Figure 5.15a).

Hot-electron temperature K_α-radiation is generated by the interaction of the hot electrons with the inner-bounded electrons of silicon atoms. The kinetic energy of the hot electrons is determined by the cross-section for K-shell ionization by electrons [154, 155]. Assuming a Maxwell velocity distribution of the electrons, the spectral energy density $w(E, T_h)$ for Bremsstrahlung radiation is given by

$$w(E, T_h) = \alpha \cdot (T_h)^{0.5} \exp\left(-\frac{E}{T_h}\right) , \tag{5.12}$$

where E is the photon energy, α is a generalized plasma parameter, and T_h is the temperature of hot-electrons [156].

When X-ray radiation is generated only by Bremsstrahlung, the measured energy of the X-rays with the photo-diodes w_{PD} is proportional to

$$w_{\mathrm{PD}} = \int_0^\infty T_i(E)w(E, T_h)dE = S_i , \quad i = 1, 2 \tag{5.13}$$

where $T_i(E)$ is the transmittance of the ith filter. The ratio of the measured energies S_1 and S_2 for two different filters T_1 and T_2 is independent on α

$$r(T_h) = \frac{S_1}{S_2} = A \cdot \frac{\int_0^\infty T_2(E)\exp(-E/T_h)dE}{\int_0^\infty T_1(E)\exp(-E/T_h)dE} , \tag{5.14}$$

where $A = 1.66$ is a calibration factor for the two photo-diodes.

The electron temperature T_h can be calculated by solving numerically the right side of Eq. (5.14) for the measured ratio r. Because of the assumed Maxwell velocity distribution of the electrons, the mean energy of the electrons can be calculated by

$\langle E \rangle = \frac{3}{2} T_h$, which is in a wide range of focal positions constant ($\langle E \rangle \approx 8\,\text{keV}$) and increases only at large focal positions (Figure 5.15b). At maximal relative number of Si-K_α-photons the hot-electron temperature amounts to $T_h = 7.7\,\text{keV}$ in accordance with literature [157].

Using double pulses, varying the ratio of the energies of the pre- and main-pulse and focal position, the number of Si-K_α-photons, compared to single-pulse ones, has been increased. The resonant absorption of laser radiation with the plasma electrons can be improved, for example, by coupling the main-pulse 36 ps after the pre-pulse with a pulse energy ratio of 65%. Also, by changing the focal position by about 8 μm thereby increasing the incident angle, the absorption of the laser radiation in the plasma at the critical electron density n_e^{crit} can be improved. Compared to single pulses, using double pulses the relative of Si-K_α-photons can be increased by a factor of four.

5.2
Imaging Detection

Imaging pump and probe metrology is realized using the probe radiation to spatially excite or transilluminate a sample. The first method is used to visualize processes, which would not be detectable by transillumination. For example the resonant-absorption photography enables the detection of vapor dynamics by exciting the atoms in the vapor resonantly and detection of the optical emission time-resolved by a CCD [32]. In general the second method by transillumination or reflection is the most common technique for pump and probe imaging. Imaging can be subdivided into

- non-coherent (Section 5.2.1), and
- coherent methods (Section 5.2.2 and Figure 5.2).

In Section 5.2.1, examples of non-coherent methods for the realization of time-resolved micro-shadowgraphy and quantitative phase microscopy are given. Exemplary for coherent methods, time-resolved speckle-microscopy and Nomarsky-microscopy are presented. Non-coherent imaging is shown using a very common technique: micro-shadowgraphy (Section 5.2.1.1). Also, a novel method, transient quantitative phase microscopy, will be presented (Section 5.2.1.2). In detecting quantitative phase information, highly useful information for ultra-fast engineering is extracted for investigation of laser-induced plasmas, changes in geometries of micro-structures in three dimensions, and refractive index changes in glasses. Coherent imaging is exemplary shown by micro-Mach–Zehnder interferometry (Section 5.2.2.1), by speckle, and as well by Nomarsky microscopy (Sections 5.2.2.2 and 5.2.2.3).

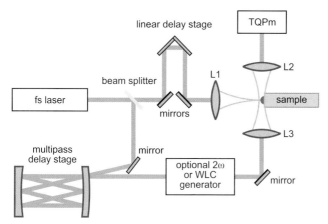

Fig. 5.16 Schematic of femtosecond pump and probe shadowgraphy [134].

5.2.1
Non-coherent Methods

5.2.1.1 Shadowgraphy

Principle and setup Time-resolved shadowgraphy experiments were performed using a femtosecond time-resolved pump-probe setup (Figure 5.16) to image the emission of particles, vapor, and melt by time-resolved shadowgraphy. A femtosecond CPA-laser system (Thales Concerto) operating in single-pulse mode at the wavelength λ = 820 nm has been used to generate the pump (processing) laser radiation, with a pulse duration t_p = 80 fs (FWHM), nearly Gaussian beam profile, and pulse energies $E_p \leq 1.5$ mJ. The laser radiation is focused by a microscope objective L1 (Olympus MSPlan20, NA = 0.4). The calculated spot size for a 20 × magnification features $2w_0 \approx 10$ μm. The applied pulse energy was set to 100 times above the ablation threshold for each material under investigation. After a beam splitter, the laser radiation passes a multi-pass optical delay stage and a 2ω or white-light continuum (WLC) generator (Section 5.1.1.2), forming a probe beam (L2, L3) that is perpendicular to the optical axis of the pump laser radiation. By adjusting the multi-pass delay stage (Section 4.4.1.2) on an optical rail and by use of an additional linear delay stage for the pump laser radiation, the optical paths of the pump and probe beams can be continuously varied. So the time difference between the initial pump pulse and the sample illumination is changed.

Polished Al or Cu samples with optical quality featuring high chemical purity (10 × 10 × 1 mm^3, $p > 99.9\%$) can be flexibly moved by three high-precision stages (PI). Shadowgraphs have been taken by photography using a CCD camera (Baumer Optronic arc4000c), using a microscope objective L2 (Olympus MS-Plan50, NA = 0.55) with the overall spatial resolution of 1.5 μm. Spatial transmittance images can be obtained by referring the measured shadowgraph images to the background images. This experimental setup enables setting time delays up

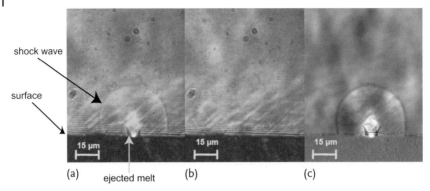

shock wave

surface

15 µm

15 µm

15 µm

(a)   ejected melt   (b)   (c)

Fig. 5.17 Melt ejection and shock wave formation after irradiation of Cu – pump laser radiation enters the system from the top ($t_{\text{delay}} = 14.6\,\text{ns}$, $F = 1.3\,\text{J/cm}^2$, $\lambda_{\text{pump}} = 820\,\text{nm}$, $t_p = 80\,\text{fs}$) [134].

to $1.2\,\mu\text{s}$ with a pulse duration $t_p \approx 100\,\text{fs}$, which can be used for observation of transient processes on surfaces.

A shadowgraph image (Figure 5.17a) recorded by a CCD camera after a single-pulse interaction with copper features an expanding hemispherical shock wave, a gas-like plume, and material droplets. In addition to these characteristic phenomena, during the femtosecond laser ablation of copper and aluminium, a jet expanding vertically upward was observed at larger delay times. Interference patterns resulting from the light scattering on the sample edges can be eliminated by referring the image with the laser-induced process (Figure 5.17a) to the background intensity without the laser-induced process (Figure 5.17b). The calculated spatial transmittance image (Figure 5.17c) provides larger contrast and inhibition of artifacts.

5.2.1.2 Transient Quantitative Phase Microscopy (TQPm)

Principle and setup Interferometry with monochromatic radiation is only suitable for dynamical detection of processes, when the quantitative phase change is $< \pi$ [33]. A novel microscopic method for time-resolved phase measurement with phase changes $> \pi$ using the commercial software QPm[38] has been developed, called transient quantitative phase microscopy (TQPm) [140], and adopts conventional bright-field transmitting or reflected light microscopy without additional optical components. QPm calculates the phase information of an object by using three images of the object taken by a CCD camera at three different object planes of the microscope. This is achieved by moving the object and taking images sequentially (Appendix B.2). Some dynamical non-reproducible processes, however, like melt dynamics cannot be detected multiple times, because the information of each picture taken varies. In order to overcome this, QPm has been improved, by taking multiple pictures simultaneously (Figure 5.18a). Unlike QPm, the developed

38) www.iatia.com.au

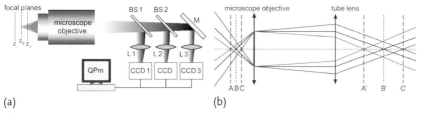

Fig. 5.18 Schemes of transient quantitative phase microscopy (TQPm) (a) and of the optical path of object planes *A, B, C,* and different image planes *A', B', C'* (b) [140].

TQPm setup combines three CCD cameras with a commercial microscope (LEICA DML), which enables synchronous image detection and thus time-resolved quantitative phase microscopy.

An adapter for these three CCD cameras has been designed to be connected on a Leica DM/LM microscope. The radiation is subdivided into three approximately equal intense beams (beam splitter 1: reflectivity 30% and transmissivity 70%, beam splitter 2: reflectivity 50% and transmissivity 50%). The object plane is imaged by an objective and the tube lens of the microscope. Different object planes *A, B* and *C* displaced by Δz are imaged on different image planes *A', B'* and *C'* (Figure 5.18b).

For the investigations the objectives Olympus ULWD MSPlan $50 \times /0.55$, Leica N Plan L $20 \times /0.4$, Leica HC PL Fluotar $50 \times /0.8$, and Leica HC PL Fluotar $63 \times /0.7$ were used.

The procedures for preprocessing of three images for the phase calculation include a calibration and an experimental routine (Figure 5.19). Detecting different planes with three CCD cameras results in images with different magnification and illumination intensity. This has to be adapted to obtain the same dimension and illumination of the images, identically to using one CCD camera and displacing the object with standard QPm. In order to compensate for this the "Intensity Module" performs an intensity transformation by taking three background images of three objects planes and calibrating them with the three images (Figure 5.19).

Characterization of TQPm   The CCD cameras have to be aligned orthogonally to the optical axis in order to ensure phase measurements precision. Image distortions have been detected for slightly misaligned CCD cameras. Mechanical imperfections of the adapter and limited alignment precision result in images taken by the CCD cameras being rotated and translated. Because of the varying magnification in different image planes, differences in scales and illumination intensities occur. Sharp images of an object taken with a conventional microscope are almost free of aberrations, because the microscope optics are well-corrected at the working point. By displacing the focal position, for example, by $\Delta z = \pm 10\,\mu m$, the resulting images are consequently subjected to aberrations, which can reduce the resolution of the phase measurements. The effects of varying magnifications and possibly misaligned optics are corrected by the "Geometrical Module" of the cali-

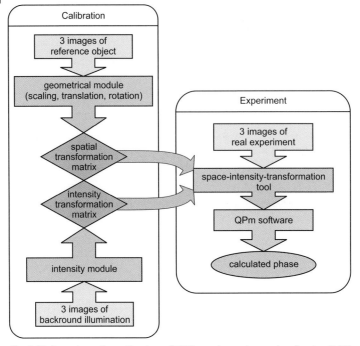

Fig. 5.19 Procedures for calibration of TQPm and experimental realization [140].

bration routine, including rotation, translation, and scaling of the involved images (Figure 5.19). A calibration plate representing crossing 2 µm wide lines has been generated by laser-induced ablation on a gold-coated glass substrate. In the first reference images of the calibration plate are taken with only one CCD at three different object planes by moving the object manually the desired Δz. This triplet of images refers to the standard QPm procedure. In the second step the images of the calibration plane are taken by three CCDs, which are placed to detect the displaced image planes corresponding to step one. An algorithm programmed in Matlab compares the reference image of each defocused plane with the corresponding image taken by the TQPm setup and adjusts them by a projective transformation. This represents a vector space transformation preserving geometrical straightness and featuring rotation, translation, and rescaling of the images. Four image points are sufficient to define a projective transformation unequivocally. Distortions induced by non-orthogonal aligned CCD concerning the optical axis and image aberrations are not compensated by projective transformation.

In order to compensate for illumination intensity variation, the "Intensity Module" performs an intensity transformation by taking three images using both standard QPm and TQPm experimental setups. Afterwards a similar procedure to the geometrical transformation is adopted without an object. Firstly, reference images at three different object planes are taken by moving the microscope condenser (Leica $NA = 0.1$–0.9) without an object a desired Δz with one CCD. In the second step

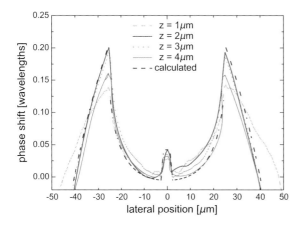

Fig. 5.20 Phase shift of a fiber as function of the displacement for core and cladding (wavelength λ = 550 nm, Leica 50 × /0.8) [140].

the images are taken by three CCDs. An algorithm programmed in Matlab compares the reference image of each defocused plane with the corresponding image taken by the TQPm setup and adjusts the image intensities by adding an appropriate offset.

The resulting geometrical and intensity transformations are described by the 3×3 spatial transformation matrix and the $H \times W$ (H and W image height and width) intensity transformation matrix respectively (Figure 5.19), which are thereafter used for image preprocessing by the experimental routine. Three images simultaneously taken of a real experiment are then preprocessed by the "Space-Intensity-Transformation tool" delivering input data for the QPm software.

The phase shift, defined as optical path per wavelength, has been measured for a commercial fiber (3M F-SN-3224 immersed in glycerin) with the Leica 50 × /0.8 microscope objective (Figure 5.20). A loss in the phase information results when the displacement Δz deviates from the depth of focus of the adopted objective considerably. The expected phase shift is calculated by the optical path length ns/λ, using the refractive index n of the fiber core, claddings and glycerin, and the corresponding transmitted geometrical path (Table 5.1).

The calculated phase shift fits well to the measurement at Δz = 2 µm, which corresponds to the objectives depth of focus. The phase shift at the fiber core and

Table 5.1 Refractive index and diameter of fiber and glycerin.

	Core	Cladding 1	Cladding 2	Glycerin
Refr. index	1.4580	1.4530	1.4570	1.449
Diameter	4 µm	50 µm	150 µm	–

at the interface cladding 1–cladding 2 (lateral position ±25 μm) is maximum. On-line pump-probe control of welding using TQPm is demonstrated in Section 6.4.

5.2.2
Coherent Methods

Coherent methods for imaging are useful for the detection of objects characterized by small absorptivities, like gases, vapor or organic particulates, generally called phase objects. Even the most common technique is the Michelson interferome-try, a more stable setup, the Mach–Zehnder micro-interferometer is used in prac-tice [158]. Additionally to this technique, speckle and Nomarsky microscopy will be presented.

5.2.2.1 Mach–Zehnder Micro-Interferometry
Principle and setup The Mach–Zehnder interferometer is a further development of the Jamin interferometer first developed in 1892. The principle of this interfer-ometer is the "round the square" system (Figure 5.21) [159]. The radiation from the source (L) is divided by the beam splitter (T1) in two separated beams, each re-flected by the mirrors S1 and S2 and recombined by the second beam splitter (T2). Thereby the radiation of both beams can interfere.

One beam represents the measurement beam, the other the reference beam. The separation length of the two beams is predefined and is an advantage com-pared to the spatially limited Jamin interferometer. A commercially available micro-interferometer based on the Horn-design was built in 1960 by Leitz (Figure 5.21).

Using polychromatic radiation for interferometry the fringes observable in the interferometer are colored (Figure 5.22). The measurement of the optical retarda-tion is achieved by measuring the displacement of the fringes at one color (or using monochromatic radiation). The optical retardation is calculated by

$$\Gamma = (n_0 - n_m)d \tag{5.15}$$

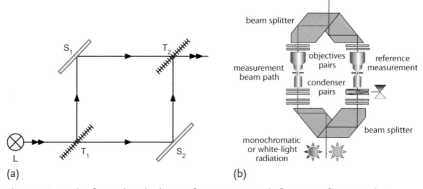

Fig. 5.21 Principle of a Mach–Zehnder interferometer (a), and of Leitz interferometer (b).

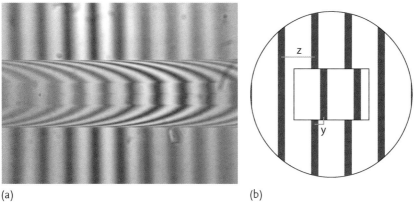

Fig. 5.22 Colored interference fringes of an optical fiber (a) and reconstruction of the fringe displacement (b).

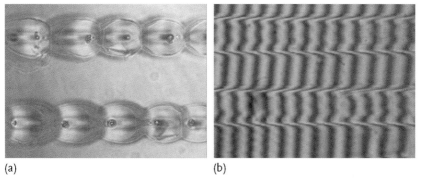

Fig. 5.23 Colored interferogram of a waveguide at 0° interpenetration angle (a) and monochrome interference fringes of an optical fiber (b).

with n_0 the refractive index of the object, n_m the refractive index of the surrounding medium, and d the thickness of the object. With given distance between neighbor fringes z, the value of the fringe displacement y and the wavelength λ (Figure 5.22b), the refractive index of an object can be calculated

$$\Gamma = \lambda \frac{y}{z}. \tag{5.16}$$

The width of the fringes depends on the interpenetration angle between the beams of the two arms of the interferometer. At an interpenetration angle of 0° the object is detected with a homogeneous background each color representing a different optical retardation (Figure 5.23). Large optical retardations $y > z$ are difficult to be evaluated using monochromatic sources (Figure 5.23). An optical retardation induced by a phase object is detected using polychromatic radiation allowing one to distinguish different colors and so enabling the detection of fringe displacements $y > z$ representing a large optical retardation (Figure 5.23).

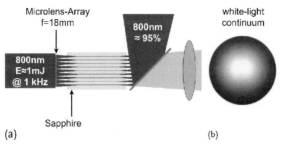

(a)   (b)

Fig. 5.24 Scheme of white-light continuum generated by laser radiation focused into sapphire with a micro-lens array (a) and detected homogeneous intensity distribution of WLC (b).

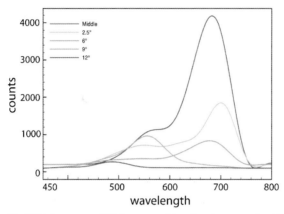

Fig. 5.25 Spectrum of WLC generated by micro-lens array for different angle.

Characterization Time-resolved Mach–Zehnder micro-interferometry has the advantage of using laser radiation to posses a larger spatial coherence and, by using a white-light continuum, to having unequivocal dependence on the measured phase for values $> \pi$. The temporal resolution should be about the pulse duration of the fundamental ultra-fast laser radiation, even when the spectral components are displaced by chirping (see Section 5.1.1.2).

The spatial intensity distribution has to be homogeneous and free of filaments, especially for the evaluation of the fringes. In order to fulfill this requirement ultra-fast white-light radiation is generated by focusing the fundamental radiation into sapphire. As shown in Section 5.1.1.2 white-light radiation is generated with only one filament at small average power. Generating simultaneously many such filaments using a micro-array lens results in a homogeneous spatially intensity-distributed ultra-fast white-light radiation (Figure 5.24). The measured angular dependence of the spectral distribution given in Figure 5.25 rises from the non-linear processes (see Section 3.1.2.3) and has to be taken into account for the experiment.

Also used at the fundamental wavelength or SHG of the ultra-fast laser radiation, this pump and probe technique is a very precise tool for time-resolved phase measurements, with the drawback of the ambiguity of multiple-π phase values due to monochromatic radiation. In order to get reliable phase measurements the objectives (Figure 5.21) should feature the same phase distortion.

5.2.2.2 Speckle Microscopy

Speckle microscopy is a method exploiting the spatial coherence of the laser radiation. Speckles are generated when coherent radiation is reflected and diffracted at a rough surface. Interference between incoming, reflected, and diffracted radiation takes place when the spatial coherence (Section 4.1.3) is larger then the lateral distance of the elevations of the surface (= average roughness). A heterogeneous spatial intensity distribution results (Figure 5.26a).

Speckles which are directly detectable are called "objective speckles" whereas "subjective speckles" need some kind of instrument in order to be imaged. Subjective speckles are mostly used in metrology and will be further discussed (Figure 5.26b). The dimensions of a speckle

$$\sigma_{\text{sp}} = 1.22\lambda(1 + M)\frac{f}{D} \tag{5.17}$$

are a statistical average defined by the distance between neighboring maxima and minima of the intensity distribution of the speckles. M defines the magnification, f the focal length, and D the diameter of the objective [160]. Due to the large coherence of laser radiation the visibility speckles is large, Eq. (4.12).

A diffusing screen with an optically rough surface made by etching the surface is also used for the generation of speckles. An alternative speckle generator can be implemented by etching the end-surface of a multi-mode fiber. In far-field the spatial intensity distribution of the speckle is circular and Gaussian (Figure 5.26a).

Object density changes in the image plane of a microscope are detected by displaced speckles not located in the image plane.

A diffuser with the surface topology described by the amplitude function $U_D(\xi, \eta)$, and illuminated by coherent laser radiation, generates a speckles pattern with the intensity distribution $I(\vec{r})$. A photographic plate is illuminated by

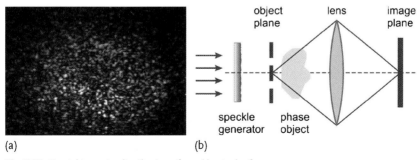

(a) (b)

Fig. 5.26 Spatial intensity distribution of speckles in the far-field (a) and scheme of speckle generator (b) (λ = 532 nm, t_p = 28 ps) [141, 144].

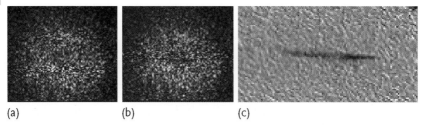

(a)   (b)   (c)

Fig. 5.27 Speckle pattern without (a), with a modification in glass (b), and calculated intensity by Eq. (5.18) (c) [141, 144].

a speckle pattern $I(\vec{r})$ and illuminated a second time by the same speckle pattern translated by $\boldsymbol{\Delta} = \vec{i} \cdot \Delta_x + \vec{j} \cdot \Delta_y$ with $\vec{i}$ and $\vec{j}$ the unit vectors for the translation. Developing the photographic plate and irradiating it with laser radiation exhibits Young's interference pattern, with a strip distance $|\boldsymbol{\Delta}|$ being aligned orthogonally to $\boldsymbol{\Delta}$.

Superimposing speckle patterns of the clear path and of a transparent object is imaged by detecting, first, the speckle pattern without the object and, secondly, with the object (Figure 5.27). A refractive index change induced in a gas or a glass, being proportional to the pressure change, displaces the speckles.

Using the speckle microscopy technique the refractive index change and the transmission change of an object are detected. Laser radiation of a diode-pumped regenerative amplified Nd:YAG laser system (pulse duration $t_p = 38$ ps, $\lambda = 1064$ nm) has been adopted [161]. The image plane of the speckle pattern has been positioned in the image plane of the object not offering a quantitative evaluation of the refractive index. Speckle microscopy becomes sensitive for the deflection angle by positioning the speckles plane into the image plane. Refractive index changes of the object do not displace the speckles. The transmittance change of the overall image is calculated by

$$T = \frac{\sum_i a_i^{\mathrm{Mes}}}{\sum_i a_i^{\mathrm{Ref}}} .$$ (5.18)

In this case the intensity values of each pixel a_i of the measurement image and the reference image are summarized.

The phase information in the image is conserved resulting in a definite phase relation between the speckles and enabling the measurement of refractive index changes by phase change measurement. For example, in the case of a change in the phase of a speckle the neighbor speckles are also affected. A summarized phase change of an image can be calculated by summarizing the transmission change of each speckle

$$\delta \Phi = \sum_i \frac{a_i^{\mathrm{Mes}}}{a_i^{\mathrm{Ref}}} .$$ (5.19)

Because the positions of the speckles are fixed, this calculation is carried out using the intensity values of the pixels of the measurement and reference image.

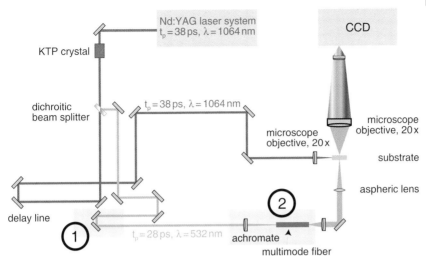

Fig. 5.28 Scheme of the setup for speckle microscopy (1. delay line, 2. speckle generator).

Combined with a transillumination microscope, the measurement and reference images are taken consecutively by photography using a CCD camera [161]. As a speckle generator frequency-doubled Nd:YAG laser radiation ($\lambda = 532$ nm, $t_p = 28$ ps) has been coupled by an achromatic objective ($NA = 0.4$) into a multimode quartz fiber (diameter 300 μm) (Figure 5.28). The probe pulse duration is increased by multiple reflections in the fiber

$$t_p^{\text{speckles}} \approx t_p \cdot \frac{L}{\sqrt{1 - NA^2}} \tag{5.20}$$

to $t_p^{\text{speckles}} = 31$ ps. The fiber end is roughened by etching. The speckle pattern is imaged by an achromatic objective onto the object plane. A microscope objective (Olympus ULWD 20 ×) images the object plane on a CCD camera. The resulting image is calculated by dividing each pixel of the measurement image through the reference image. The noise has been reduced by averaging three images.

5.2.2.3 Nomarsky Microscopy

Principle and setup Nomarsky microscopy (also called differential interference contrast microscopy) was developed 40 years ago for the observation of phase objects [162]. This technique can overcome the cell-poisoning coloring of objects to emphasize them.

Nomarsky microscopy uses spectrally broad radiation $\Delta\lambda \approx 10$ nm, for example, radiation of a halogen bulb. The radiation is linearly polarized by a polarizer, and through a Wollaston prism the radiation is separated spatially into two orthogonal polarized components (Figure 5.29). The spatial displacement represents several μm. The two radiation components are collimated by the condenser and experi-

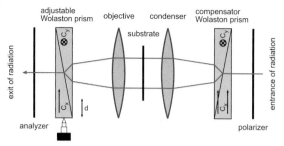

Fig. 5.29 Scheme of the principle of Nomarsky microscopy [163].

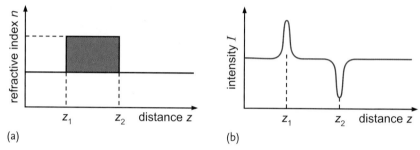

(a)   (b)

Fig. 5.30 Principle of Nomarsky microscopy: a step-like refractive index n change (a) and resulting intensity distribution
(b)[141, 144, 164].

ence phase shifted in the substrate. Afterwards the two radiation components are
recombined by an adjustable Wollaston prism.

Due to the spatial separation of the linear polarized radiation components, different locations within the beam exhibits a phase shift. By selecting the polarizer
the recombined radiation components interfere. The interference takes place only
in the Wollaston prism. Interference occurs when the spatial separation is smaller
than the spatial coherence of the radiation. The maximal separation is defined as
the coherence length of the used radiation.

The dimensions of the contrast change in the image have to be on the order
of the resolution limit of the microscope and the spatial coherence has to be as
large too. For example, a light source used for Nomarsky microscopy based on
thermal emission exhibits a broad spectral distribution with a coherence length of
several μm. Using laser radiation the temporal coherence length is given by the
pulse duration below $t_p = 100\,\text{fs}$ resulting in a spatial coherence of about 30 μm.

A one-dimensional step-like change of refractive index results, observed by Nomarsky microscopy, as the derivation of the refractive index ∇n exhibiting a intensity increase and decrease around the positions of refractive index changes (Figure 5.30). Changes in the refractive index of phase objects are reproduced by Nomarsky microscopy as high-contrast images.

Time-resolved Nomarsky microscopy is a pump and probe technique using ultra-
fast white-light continuum WLC (Figure 5.31). The WLC is generated by self-phase

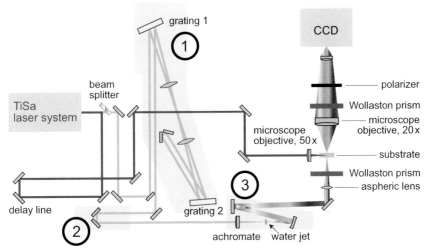

Fig. 5.31 Setup of time-resolved Nomarsky microscopy
(1. Zero-compressor, 2. Delay-line, 3. WLC generator)
[141, 143, 144, 164].

modulation of ultra-fast laser radiation in a dielectric (Section 5.1.1.2) and exhibits small pulse durations ($t_p^{WLK} \leq 3.5$ ps). Each spectral component features a pulse duration comparable to the one of the generation laser radiation ($t_p^{\lambda} \leq 100$ fs) resulting in a coherence length comparable to thermal sources. The WLC is linearly polarized. The images are detected by photography using a color CCD camera (Arc4000c, Baumer Optronic, 1300×1030 Pixel^2) and a modified Normarski-microscope (DMLP, Leica). The color detection is established by a Bayer matrix on the CCD chip with each color filter (RGB) having 50 nm bandwidth. The temporal resolution of the Normarski photography for each color of the CCD camera can be calculated to 1 ps using the chirp of the WLC (Section 5.1.1.2).

Characterization The filaments of the WLC are detected by CCD as a speckle pattern (Figure 5.5). The description of this chaotic intensity distribution is difficult.

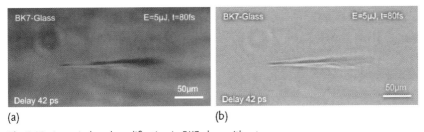

(a)   (b)

Fig. 5.32 Laser-induced modification in BK7 glass without filter (a) and a high-pass filter (b) (Nomarsky microscopy, $\lambda = 810$ nm, $t_p = 80$ fs, $I = 10$ PW/cm^2) [141, 143, 144, 164].

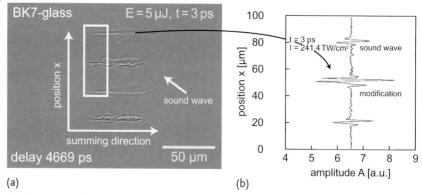

(a)   (b)

Fig. 5.33 Laser-induced modification in BK7 glass (a) and spatial distribution of the sound wave (b) (Nomarsky microscopy, $\lambda = 810$ nm, $t_p = 3$ ps, $I = 241$ TW/cm^2) [141, 143, 144, 164].

Image processing of the Nomarsky images (analyze radius five pixels) is adopted using a high-pass filter to homogenize the intensity distribution. Transmission changes are suppressed by this processing enhancing only the phase changes (Figure 5.32). Only strong transmission reductions, induced, for example, by plasma absorption, can be detected.

A signal proportional to the amplitude of the sound waves generated by laser-induced modification of BK7 glass can be detected as a function of the position (Figure 5.33). The spatial distribution of the sound waves can be extracted from Nomarsky images. The presented techniques enable the detection of different processes with femtosecond resolution during the processing of materials as shown in Section 6.

6
Applications of Pump and Probe Metrology

Ultra-fast laser radiation is used in industrial applications to modify, melt, or ablate matter. Related to these processes different processing, like coloring, welding or structuring, can be deduced (Figure 6.1). The processing of metals by drilling and structuring are investigated using the imaging pump and probe techniques described in the Section 5.2. The first processing is detected by shadowgraphy, whereas structuring is investigated by speckle microscopy.

Glass is processed using ultra-fast laser radiation for marking and welding applications. Marking glass is investigated by transient absorption spectroscopy, a non-imaging technique (see Section 5.1), and by the imaging technique Nomarsky microscopy. The optical phase change induced during the welding of glass is detected by a novel imaging technique, quantitative phase microscopy.

The pump and probe techniques described in the following: drilling, structuring, marking and welding, are used to get a deeper process understanding with the aim to develop industrial reliable processing.

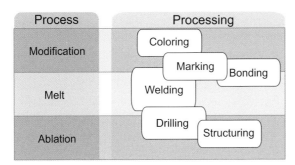

Fig. 6.1 Ultra-fast processes and derived processing.

Ultra-fast Material Metrology. Alexander Horn
Copyright © 2009 WILEY-VCH Verlag GmbH & Co. KGaA, Weinheim
ISBN: 978-3-527-40887-0

6.1
Drilling of Metals

6.1.1
Introduction

The automotive, aeronautical and energy industries would like to apply new technologies to reduce fuel consumption because of the harmful exhaust gases. One strategy is increasing the hole density of turbine blades used in gas turbines by reduction of the diameter of the cooling holes, increasing at the same time the number of holes. Also, for injection nozzles used in diesel engines a more effective nebulization is achieved by more reproducible holes with small diameter. Drilling holes by laser radiation is today industrially established for hole diameters greater then 100 μm using millisecond and microsecond pulsed Nd:YAG lasers (Figure 6.2). The generation of holes with diameters < 100 μm with large reproducibility and productivity is still not fulfilled. One attempt is the application of high energy ultra-fast laser radiation [134, 135, 165].

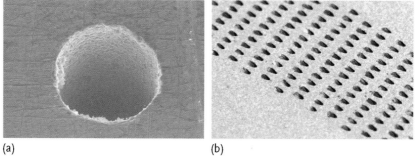

(a)   (b)

Fig. 6.2 Drilling in steel [166] (a) and contoured holes in single crystal alloy by pulsed laser radiation [167] (b).

In the following section the single-pulse ablation of metals by high-intensity femtosecond laser radiation is detected by imaging pump and probe methods, like shadowgraphy and quantitative phase microscopy. Even with one pulse melting is induced.

6.1.2
Measurement of Ejected Plasma, Vapor and Melt

Time scales of nanoseconds for evaporation and of microseconds for "normal boiling" after heterogeneous nucleation have been reported [168, 169]. Modeling [170] performed for metals above the metal boiling temperature, however, describes the removal of only several atom layers in 100 ns. Therefore, this process seems to be negligible for timescales < 1 ns. Concerning boiling effects, heterogeneous nucle-

 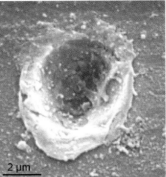

Fig. 6.3 Surface structures induced in Cu (a) and Al (b) by single pulse ultra-fast laser radiation (SEM, λ = 820 nm, $F = 0.3 - 0.4$ J/cm^2) [134].

ation in the liquid is involved. For metals at twice the melting temperature result in diffusion distances < 1 nm in 100 ns. Therefore, normal boiling is also not expected to yield in a significant contribution to the ablation for the time scale < 1 µs. Experiments on GaAs, a semiconductor, below the plasma formation threshold [112], however, show bubble-like structures in the liquid. These bubbles appear about 20 ns after the ablation pulse and grow linearly in time.

Qualitative differences of the performed surface modifications result in metals such as copper and aluminum (Figure 6.3). The surface morphology of copper reveals rosette-like modifications. Re-solidified melt droplets and material jets on sub-micrometer size scale are observed in the distance of some 5–10 µm from the center of the ablated region. Slightly elliptical contours of the produced craters correspond to the spatial beam profile of the laser radiation used. In contrast to that, the crater contours on aluminum targets retain sharp edges. A measure for the heat conductivity of metals and the energy transfer is given by the electron–phonon coupling constant resulting in a complex hydrodynamic movement of the melt. Besides the similar optical properties of copper and aluminum, the smaller electron–phonon coupling constant for copper (γ_{Cu} = 10 × 10^{16} W m^{-3} K^{-1} and γ_{Al} = 4.1 × 10^{16} W m^{-3} K^{-1}) result in a larger amount of melt when irradiating copper with ultrashort laser radiation.

The dynamics of the melt formation and material ablation for single pulse irradiation at larger pump-to-probe elapsed times up to τ = 1.8 µs (Figure 6.4) reveal a variety of phenomena, like plasma formation, ejection of material droplets and liquid jets [134, 165, 171]. The phenomenological development of the ablation dynamics on the time delay ranges from τ = 3.0 ns to τ = 1.04 µs to be conditionally categorized into three characteristic time regions.

1. The first region features photo-induced emission, formation and expansion of highly pressurized, heated material, combined with the initial formation of shock waves (Figure 6.4a and b). At the time delay τ = 49 ns (Figure 6.4a) an expanding shock wave and vapor plume in gas has been detected and can

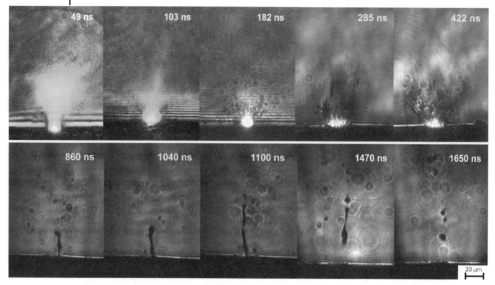

Fig. 6.4 Shock wave, plasma and melt emitted from Al as function of time delay (Shadowgraphy, $F = 1.8\,\text{J/cm}^2$, $\lambda_{\text{pump}} = 820\,\text{nm}$, $t_p = 80\,\text{fs}$, WLC) [171].

be described by a combined model developed for nanosecond ablation [168, 169].

2. At approximately $\tau = 200\,\text{ns}$ time delay, in the second region (Figure 6.4c and d), the ablated material is ejected explosively. Observation of ejected droplets and clusters with diameters $\approx 1 - 3\,\mu\text{m}$ can be detected for Cu and Fe, and is related to the nucleation and boiling effects [168–170].

3. During the third characteristic time region at time delays $\tau > 700\,\text{ns}$ (Figure 6.4e and f), a jet of molten material expanding into the gas atmosphere at a velocity of $\approx 100\,\text{m/s}$ has been observed on aluminum targets. The estimated height of the ejected structures averages about $50\,\mu\text{m}$ at the delay $\tau = 1.04\,\mu\text{s}$. Results on nano-jets and micro-bumps formation on gold films performed at comparable fluences [112] can be compared qualitatively to the transient melt ejection, with respect to different absorption and heat conduction in bulk and the thin film. Because of the similar thermal properties and similar small electron–phonon coupling constants of aluminum and gold ($\mu_{\text{Al}} = 4.1 \times 10^{16}\,\text{W m}^{-3}\,\text{K}^{-1}$ and $\mu_{\text{Au}} = 2.3 \times 10^{16}\,\text{W m}^{-3}\,\text{K}^{-1}$), melt is formed. Unlike the experiments of Korte *et al.* [172], the experiments were carried out ablating metals with laser radiation well above ablation threshold: the generated melt structures mostly do not result in permanent nano-jets.

In order to obtain the information from a definite volume ΔV in the interaction area of the ejected particles, transient quantitative phase microscopy (TQPm) based

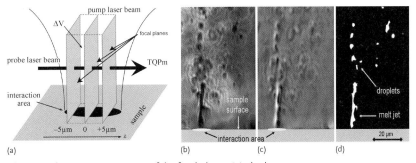

Fig. 6.5 Schematic arrangement of the focal planes (a) shadowgraph image (b), quantitative phase (c), and TQPm results (d), showing melt ejection in aluminum (τ = 1.45 μs, E_p = 105 μJ) [134].

on IATIA QPm technology (see Section 5.2.1.2) has been adopted (Figure 6.5) (Section 5.2.1.2) [171].

Due to the small dimension of the ablated material, such as ejection of vapor, melt droplets, and jets, whose dimensions often do not exceed several micrometers, shadowgraphy images often lack depth of sharpness. Thus, corresponding volume information, phase, object thickness, consistency, and geometry of molten material, cannot be detected reliably. By TQPm the image contours within a volume ΔV in the interaction area with a predefined thickness 10 μm can be detected (Figure 6.5a). By adopting filters in the calculated phase by TQPm, material droplets from the entire area are clearly detected (Figure 6.5c). Ablated material volume, for example evaporated and/or molten metal, may be calculated by assuming cylindrical symmetry of the droplets contours along the surface normal [135].

6.2
Microstructuring of Metals

6.2.1
Introduction

The implementation of ultra-fast laser sources in microtechnology is advancing. Productive microstructuring of metals for the semiconductor industry needs ultra-fast laser radiation for large quality, like roughness R_a < 10 nm. For example, the texturing of large surfaces with microstructures needs high-repetition ultra-fast lasers (Figure 6.6).

Ablation of thin metal film of gold and copper has been investigated by pump and probe speckle microscopy (see Section 5.2.2.2) [161, 173, 174]. The interaction zone irradiated with ultra-fast pulses (λ = 1064 nm, t_p = 35 ps) has been investigated by frequency doubled speckle-pattern-modulated radiation. A spatial resolution of 1 μm has been achieved by using microscope objectives. The temporal resolu-

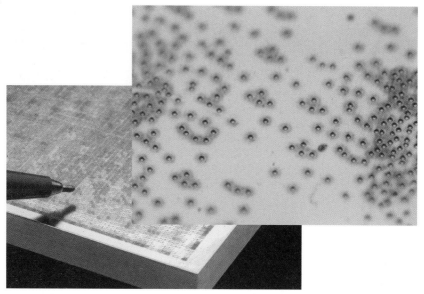

Fig. 6.6 Microstructures on steel.

tion is given by pump pulse duration of the laser source. The generation of an ablation hole, and the interaction of the vapor with the ambient gas have been investigated the first 20 ns after irradiation with one pulse observing the change of surface properties (solid, liquid or vapor) (Figure 6.7). The change in reflectivity has been detected in a collinear illumination set-up and by an orthogonal illumination the transmissivity has been detected as function of the pulse energy and delay time.

6.2.2
Detection of Plasma Dynamics

The irradiated copper surface exhibits a strong change in reflectivity 600 ps after irradiation (Figure 6.7). Three regions have been detected:
- region I is attributed to a surface plasma consisting of ionized metal,
- region II exhibits the shock front, whereas
- region III represents the ionization front.

The plasma region is characterized by strong ablation and becomes larger with increasing pulse energy. The shock wave consisting of pressurized ambient air detectable due to the induced refractive index change in air. With increasing irradiated energy the shock wave is more pronounced. The shock wave exhibits a spherical geometry, with the expansion depending nearly not on the physical properties of the ablated metal. Behind the shock wave an ionization front is developing, becoming more pronounced with increasing energy. The ionization front represents a highly

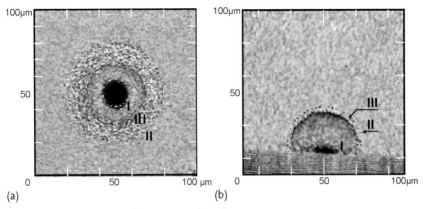

Fig. 6.7 Expanding plasma plume measured in reflectivity
(a) and in transmissivity (b) by shadowgraphy of an irradiated
copper surface, I: Plasma, II: Shock, and III: Ionization front
(λ = 1064 nm, t_p = 35 ps, $\Delta t \approx$ 600 ps) [161].

compressed gas representing the region of optical emission by a plasma. Behind
this front the gas temperature strongly increases causing ionization.

The fraction of optical energy needed for the development of the shock wave
can be calculated by the Sedov model [175], assuming the formation of a shock
wave after temporal and spatial singular energy deposition. In three dimensions
the dependence of the position of the shock front r on the energy E and time t is

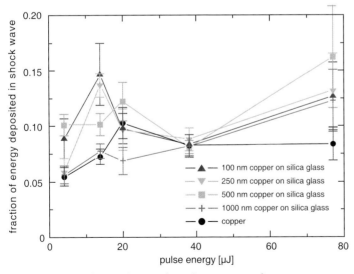

Fig. 6.8 Fraction of energy deposited into the generation of
a shock wave as function of the pulse energy (λ = 1064 nm,
t_p = 35 ps) [176].

given by

$$r = \lambda_0 \cdot t^{0.4} \left(\frac{E}{\varrho} \right)^{0.5} ,$$ (6.1)

where λ_0 represents the adiabatic coefficient of ambient air and ϱ the density of the gas. The energy fraction in the shock wave consumes about 10–20% of the pulse energy applied (Figure 6.8). This energy is lost for the ablation process. A strong absorbance has been detected during irradiation indicating that the ablation starts during the irradiation itself.

6.3
Marking of Glass

6.3.1
Introduction

Inner marking of high-quality jars is enabled by inducing microstructures in glass using laser radiation. Very precise and high-contrast writings are generated, for example, for security purposes like marking. By multi-photon ionization electron–hole pairs are formed and subsequently redox reactions of polyvalent ions are activated, in order to color glasses such as blue, mauve, yellow, red-brown and gray, which enables stress relieved marking of glass without cracking (Figure 6.9) [177, 178].

Puzzling effects of optical breakdown, avalanche ionization, multi-photon ionization or laser-induced color center formation are under investigation up to recent-

Fig. 6.9 Colored glasses by pulsed UV-laser radiation ($\lambda = 355$ nm, $t_p = 30$ ns) [177, 178].

ly [179, 180]. The interaction of laser radiation with dielectrics can be subdivided into three phases:

- absorption of radiation,
- optical emission, and
- modification of the dielectrics.

The phase shift and the absorption coefficient of MgO, SiO$_2$, and diamond have been investigated by time-resolved pump and probe spectroscopy in order to understand the processes in the photon–matter interaction during and after irradiation with pulsed laser radiation in the femtosecond regime [181, 182]. Irradiation of insulators like alkali halides results in free electrons and holes with subsequent recombination to self-trapped excitons (STEs). The recombination takes place by formation of non-bridging oxygen hole centers (NBOHCs), which decay to STEs by emission of two eV photons [183]. The recombination process to STEs of quartz is much faster than that of the alkali halides showing at low temperatures a strong luminescence for example at 2.8 eV [184, 185].

The process of electron–hole formation has been investigated here by transient absorption spectroscopy (Section 5.1.1) after excitation of fused silica with pulsed laser radiation ($t_p = 80$ fs, $\lambda = 810$ nm, $f_p = 1$ kHz).

Modifications like cracking and refractive index changes, investigated by time-resolved Nomarski-photography via the pump and probe method (Section 5.2.2.3) within two different excitation time regimes ($t_p = 80$ fs and 3 ps at $\lambda = 810$ nm), will be presented. Refractive index changes and cracking in fused silica and BK7 glass have been observed from the femtosecond to the nanosecond time regime.

6.3.2
Detection of Laser-Induced Defects

Electron–hole pairs are generated in the interaction zone, which can recombine radiatively or by charge exchange and ion transport forming self-trapped excitons (STE). STEs themselves can also recombine by optical emission. These electrons absorb radiation in the UV-VIS regime like free electrons or electrons in the conduction band of metals. The initial process of electron excitation by femtosecond laser radiation in fused silica has been observed by transient absorption spectroscopy (TAS). No conclusions about the absorption can be drawn in the time regimes 10–120 ns, because the heated region changes the refractive index strongly and deflects the radiation. Also, a TAS measurement exciting with picosecond laser pulses has not been carried out, because the glass matrix is heated during the excitation by the laser radiation.

The obtained absorption spectra feature a curve like the WLC (Section 5.1.1.2). This means that the absorption spectra are temporally and frequency shifted (Figure 6.10). An absorption band ranging from 350–750 nm, existing for a short time step (δ-function like) has been detected. The spectral bandwidth of the TAS is identical to the one of the WLC, which is mainly the same as the probe beam. The ab-

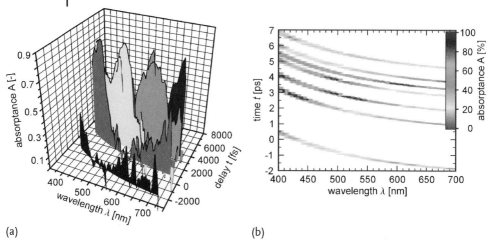

(a)   (b)

Fig. 6.10 Absorption spectrum as function of delay time in two dimensions (a) and three dimensions (b) (fused silica, $\lambda = 810\,\text{nm}$, $I = 10\,\text{PW/cm}^2$, $t_p = 80\,\text{fs}$)[144, 148].

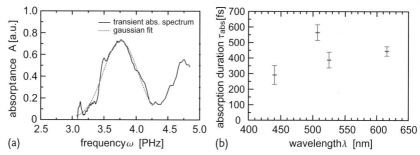

(a)   (b)

Fig. 6.11 Transient absorption spectrum for one delay time of a chirped WLC (a) and absorption duration as function of wavelength for fused silica (b) ($\lambda = 810\,\text{nm}$, $t_p = 80\,\text{fs}$) [144, 148].

sorption band changes its position according to the chirp of the WLC (Figures 5.10 and 6.10).

The absorption spectrum of fused silica excited by ultra short laser radiation exhibits an absorption band from the UV to VIS, with the absorption persisting for only a few ps (Figure 6.11a). After 8 ps no absorption band can be detected [143].

The duration of absorption is calculated as $\tau_{\text{band}} = 670\,\text{fs}$ (Figure 6.11a). As described in Section 5.1.1.2 the duration of the absorption can be calculated. Assuming an instantaneous absorption the resulting absorption duration is $\tau_{\text{abs}} = \tau_{\text{band}} - \tau_{\text{WLC}}^{\text{chirp}} \approx 300\,\text{fs}$ being independent of the wavelength (Figure 6.11b). This short absorption duration can be attributed to the formation of free electrons within a few femtoseconds [186], which recombine in the nanosecond regime or form stable defects, like STEs.

6.3.3
Detection of Refractive-Index Changes and Cracking

The interaction of pump radiation (t_p = 80 fs and 3 ps) focused tightly within BK7 glass, fused silica, or quartz 300 µm below the surface has been investigated by time-resolved Normarski-photography via the pump and probe method (Figure 5.31, Section 5.2.2.3). Orthogonal to the pump radiation, the WLC has been focused with an aspheric lens onto the substrate (beam diameter ≈ 500 µm). The radiation has been collected from the observation plane by photography using a fluorescence microscope objective (*NA* 0.3).

In the picosecond time regime the formation of filaments accompanied by a dense plasma can be observed in fused silica, quartz and BK7 glass along the beam waist. Refractive index changes along these filaments persist for some nanoseconds. Also in the first 500 fs a bright blue luminescence is observed (Figure 6.12a).

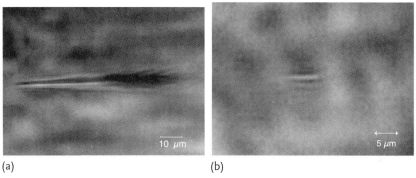

(a)   (b)

Fig. 6.12 Normarski-photography of BK7 glass 500 fs after irradiation at I = 10 PW/cm^2 (a) and I = 7 TW/cm^2 (b) (λ = 810 nm, t_p = 80 fs) [144, 148].

With increasing pulse energy, the density and lifetime of the plasma increases. In the picosecond to nanosecond time regime (10 ps to 100 ns), plasma can no longer be detected. By relaxation of the electrons or de-trapping of STEs, the glass matrix has been heated, originating in a strong refractive index change. This heated state remains unchanged for many nanoseconds. A sound wave expands from the irradiated area with a velocity of about 5–6 km/s. Cracking is observed for all the investigated materials at an intensity $I \geq 10$ PW/cm^2. For smaller intensities only refractive index changes are detected, which may be used for waveguide writing.

The glass is heated by a femtosecond laser radiation at $I < 0.18$ PW/cm^2 and cools within 40 ns (Figure 6.12b). No refractive index change is observed at all 100 ns after the exciting laser radiation.

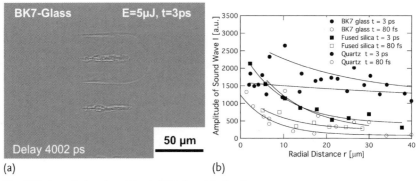

Fig. 6.13 Laser-induced cracking in BK7 glass (a) after 4 ns ($\lambda = 810\,\text{nm}$, $t_p = 3\,\text{ps}$, $I = 48.3\,\text{TW/cm}^2$) and amplitude of sound wave as function of radial distance and pulse duration for BK7, fused silica, and quartz ($E_p = 5\,\mu\text{J}$) (b) [144, 148].

During the excitation of BK7 glass with a picosecond laser pulse of high intensity, strong absorption in the irradiated region is observed. A sound wave is formed, which expands in BK7 glass with a constant velocity $v = (6.45 \pm 0.03)\,\text{km/s}$ (Figure 6.13a). The sound waves in BK7 glass expand cylindrically to the beam caustic, whereas in fused silica and quartz also spherical sound waves are formed within the interaction zone [143]. For picosecond laser excitation, an increased absorptivity by electrons is observable for about 10 ns. A refraction index change of the filaments remains observable up to 120 ns. Strong cracking is observed for all the investigated materials in this excitation regime.

The mechanical stress, measured as the amplitude of the sound wave, is larger for both kinds of glass and quartz when excited by picosecond laser radiation at the same pulse energy (Figure 6.13b). The amplitudes of the sound waves for fused silica are independent of the pulse duration and smaller than in BK7 glass. The amplitude of the sound wave for quartz excited by picosecond laser radiation decreases slowly with increasing radial distance, whereas it decreases within 20 μm when excited by femtosecond laser radiation being more pronounced for BK7 glass than for fused silica and quartz. Because of the ultrashort induced stress sound waves with a large frequency spectrum are generated, which loose energy by dispersion in the glass.

6.4
Welding of Technical Glasses and Silicon

Reliable glass–glass and glass–silicon micro-joining techniques are actually not available for joining of glasses and are realized today using techniques based on adhesive agents or interlayers. The mechanical properties and the chemical stability of the welds are not sufficient for many applications.

6.4.1
Introduction

Ultra-fast laser radiation is applied for precise microstructuring by ablation without affecting the bulk material with heat and thermo-mechanical stress [141]. Because of ultra-fast pulse duration, high-intensities are achievable at small fluences enabling multi-photon processes in dielectrics below the ablation threshold. The observed effects are irreversible refractive index change, birefringence, and electro-optical activity. These effects are not completely understood, but some approaches are given:

- A non-thermal process initiated by the generation of free electrons due to multi-photon absorption induces defects like short-living STE, long living NBOH and F-centers [187]. The density is changed by changing binding distances between silicon and oxygen atoms.
- On the other hand, a thermal process is possible by heating the glass due to the relaxation of the electron system into the phonon system [188]. The glass is heated locally [165] and large compression forces within a very small volume change the density of the glass.

Welding of glass and also of silicon has been achieved (Figure 6.14) [189–194].

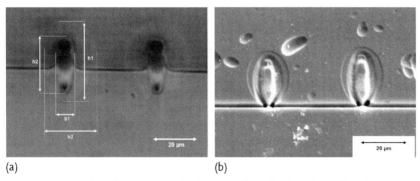

(a)   (b)

Fig. 6.14 Cross-section of welding seam in glass–glass (a) and etched glass–silicon (b).

6.4.2
Detection of Laser-Induced Melting

The joining of thin glass–glass plates by welding with femtosecond laser radiation is investigated by transient quantitative phase microscopy (TQPm) (see also Section 5.2.1.2) [140, 165, 195, 196] and Mach–Zehnder micro-interferometry (Section 5.2.2.1).

Transient quantitative phase microscopy (TQPm) Plates of technical borosilicate glass (Schott D263) with thicknesses 1 mm and 200 μm have been processed by RCA-cleaning[39] [197]. Subsequently, glass with glass has been pressed together. Joining is achieved by focusing high-repetition rate ultra-fast laser radiation (IMRA μJewel D-400, λ = 1045 nm, t_p = 350 fs) within the interface and moving the plates relative to the laser focus parallel to the interface between the plates. The processed repetition rate for glass–glass welding has been set to f_p = 0.7 MHz. The laser radiation is guided by mirrors from the laser source to a positioning stage (Kugler Microstep) and focused by a microscope objective to a diameter of 4 μm (Leica 20 ×, *NA* 0.4) to the interface of the glasses (Figure 6.15). Welding seams of D263 glass are generated moving the plates parallel to the interface at a fixed velocity (v = 60 mm/min) and focal position relative to the interface, and the phase distribution has been measured by TQPm.

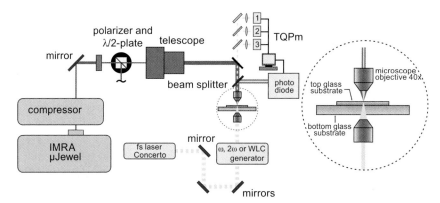

Fig. 6.15 Set-up for welding (solid line) and for Transient Quantitative Phase microscopy TQPm (dashed line) [140, 165, 196].

The optical phase within the welding seam has been detected by combining TQPm with two laser sources (Figure 6.15). For illumination of the welding area one laser source (THALES Concerto, λ = 800 nm, f_p = 80 fs) is used at 0.5 Hz repetition rate. The temporal dependence of the pump pulse (IMRA) to the illu-

39) W. Kern developed the basic procedure in
1965 while working for RCA, the Radio
Corporation of America

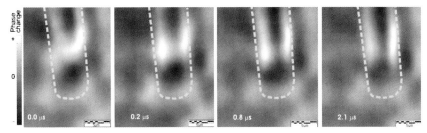

Fig. 6.16 Phase distribution of welding seam tip (dashed line) detected by TQPm for different delay times [140, 165, 196].

minating probe pulse (Concerto) is detected by a photo-diode. The laser systems are not temporally synchronized, resulting in random delays between pump and probe radiation (see Section 4.4.2.1). After measuring the data they have to be resorted. The detection of the optical plasma emission by TQPm close to the detection wavelength can be suppressed by adopting edge filters, highly transmissive at wavelengths > 750 nm.

The optical phase distribution for four delay times after pump pulse (IMRA) emphasizes an optical phase decrease within the laser interaction zone shortly after irradiation ($t = 0.0\,\mu s$). This decrease can be attributed to the generation of free electrons. Nearby the laser interaction zone an increase of the optical phase is measured which can be attributed to compressed melt. With increasing delay up to $t = 2.1\,\mu s$, the optical phase within the interaction zone increases due to cooling and re-solidification of the melt (Figure 6.16). The detection of the optical phase by TQPm is a feasible approach for process monitoring.

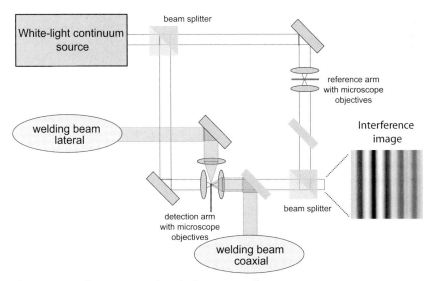

Fig. 6.17 Set-up for transient Mach–Zehnder micro-interferometry.

Transient Mach–Zehnder micro-interferometry A comparable set-up to the previous one has been adopted for phase-change detection using Mach–Zehnder micro-interferometry instead of TQPm (see Section 5.2.2.1). In this way the welding set-up has been integrated into the interferometer (Figure 6.17). Two identical interferometric-adapted 20 × microscope objectives have been chosen for the reference and the detection arm. The reference arm contains a glass substrate which induces the same optical phase shift as the sample used for detection of optical phase changes. A spatial homogeneous ultra-fast white-light continuum using IR-femtosecond laser radiation has been applied as probe radiation (see Section 5.2.2.1). The white-light continuum has been generated by THALES Concerto, $\lambda = 800$ nm, $f_p = 80$ fs at 0.5 Hz.

The optical phase within the welding seam has been detected by combining the Mach–Zehnder micro-interferometry with the welding pump laser (IMRA). The temporal dependence of the pump pulse (IMRA) to the illuminating probe pulse (WLC-Concerto) is detected by a photo-diode. The laser systems are not temporally synchronized, resulting in random delays between pump and probe radiation (see Section 4.4.2.1). After measuring the data have to be re-sorted.

The coaxial measured colored interferogram has been evaluated detecting the position of the fringes for each color and calculating the spatially resolved laser-induced phase shift (see Section 5.2.2.1 and Figure 6.18). A phase shift up to 250 nm has been detected. Knowing the dependence of the refractive index of the glass as function of the temperature dn/dT close to the melting temperature of the glass enables the calculation of the temperature distribution using the measured phase dependence.

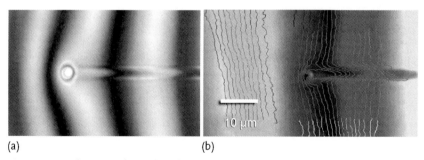

(a)   (b)

Fig. 6.18 Interferogram of coaxial welding seem in glass (a) and extracted phase shift with contour lines (b) (IMRA, 350 fs, 1 MHz, 350 nJ).

The detected optical phase shift has been evaluated on the axial position of the weld (Figure 6.19). The optical phase shift decreases in the region of the laser focus due to laser-induced defects and free electrons. Free electrons exhibit a refractive index $n < 1$. Because of heat conduction, the front weld is heated and consequently the refractive index is reduced compared to the non-irradiated region. Close behind the laser focus the refractive index is increased and larger than the non-irradiated

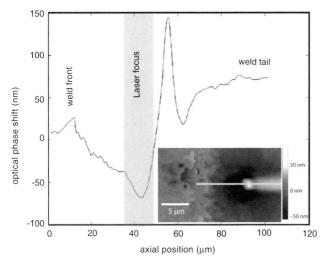

Fig. 6.19 Optical phase shift as function of the axial position in a laser-induced weld of glass (phase plot see inlet) (IMRA, 350 fs, 1 MHz, 350 nJ).

region due to highly compressed melt. The weld tail represents the re-solidified glass, which due to laser-induced stable defects and compressed glass is larger than the non-irradiated glass.

Both diagnostics enable the on-line detection of the welding process parameters like the phase shift of the glass and control of the welding process by regulating, for example, the average power of the laser radiation.

7
Perspectives for the Future

Pump and probe metrology is today mostly a scientific instrument. Perspectives for ultra-fast optical metrology are given in the following:

- Online diagnostics for quality control during processing and security applications. THz radiation, called T-rays, for example, are commercially available for security applications, like package inspection detecting metals within packaging or clothes [198]. Also for material testing, time-resolved FTIR-spectroscopy for mechanically high-stressed components, like rotating parts in engines, are becoming industrially applicable.
- Offline diagnostics for quality assurance in semiconductor industry is achieved with a picosecond ultrasonic laser sonar [199], inspecting thin-films in an integrated circuit on a semiconductor wafer by using high-repetition picosecond IR laser radiation in a pump and probe set-up. The thickness of a layer is detected by measuring the velocity of laser-induced sound-waves.
- Process understanding for processing optimization, and
- Process control as a consequence of process understanding.

In order to fulfill the transfer of ultra-fast optical metrology to industry:

1. laser sources have to become more reliable, turn-key and cheaper, and
2. the methodology for pump and probe has to become stable and fast, with small dimensions of the apparatus and reproducible into the nanometer scale serving new markets as nanoengineering.
3. The transition to industry is controlled by cost-benefit factors, in other words, this new technology is not developed unless it is commercially viable.

Perspectives in new sources and methodologies, for example, the drilling of metals (Figure 7.1), demonstrate that increasing the repetition rate of ultra-fast laser sources from 1 kHz to 1 MHz and simultaneous reduction of pulse duration and/or focus diameter by non-linear processes enable high-throughput nano-structuring. Ultra-fast optical metrology offers the investigation of the processes for the new area of processing increasing understanding and offers essential quality assurance by process control.

Ultra-fast Material Metrology. Alexander Horn
Copyright © 2009 WILEY-VCH Verlag GmbH & Co. KGaA, Weinheim
ISBN: 978-3-527-40887-0

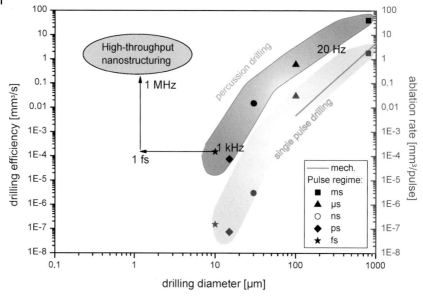

Fig. 7.1 Drilling efficiency and ablation rate versus diameter for mechanical and laser drilling by single pulse and percussion drilling with different pulse durations and extrapolation to high-throughput nano-structuring (data partly from [2]).

Perspectives for new laser sources are described in Section 7.1 and perspectives for methodology improvements in optical pump and probe metrology are given in Section 7.2.

7.1
Laser and Other Sources

Ultra-fast laser sources are becoming more reliable. On the one hand, today solid state lasers are all-diode pumped and sealed cases, thus featuring long life-time and stability. On the other hand, fiber technology for ultra-fast lasers is evolving into the larger pulse energy and power regime. The development of ultra-fast laser sources with an average power above $P = 100\,\text{W}$ at larger repetition rates $f_p \approx 1\,\text{MHz}$ are on the verge of being applicable [58, 200, 201]. Depending on the dimensions of the mirrors, adaptive optics compensating for thermally induced aberrations by high-power laser radiation may become necessary. Finally, using these high-power lasers, EUV and X-ray radiation can be generated by higher harmonic generation. Applications, like high-speed space- and time-resolved wafer inspection by pump and probe metrology can become accessible and enables interruptions of operation to be diagnosed faster.

Alternative sources like free-electron lasers, today large facilities, could be miniaturized, by micro-technologies, to become very versatile sources. The generation of

X-rays with wavelengths below 15 nm by a micro-X-FEL (X-ray free-electron laser) would revolutionize the X-ray metrology due to the variable energy adjustment necessary for medical diagnostics.

7.2
Methodology

Optical pump and probe metrology can be improved for industrial applications by increasing productivity, scanning techniques (Section 7.2.1) and reducing the apparatus complexity of temporal scanning (Section 7.2.1.2).

7.2.1
Scanning Technology

7.2.1.1 Spatial Scanning
Metrology for quality assurance has to be fast in order to be implemented into the production chains. This implies spatial scanning based technologies. Laser sources are available today at repetition rates > 1 GHz. In order to get good statistics of a measurement, but also a large operational capacity, scanning velocities in the range from $v = 100$ m/s up to $v = 1$ km/s are necessary. New strategies like electro-optics deflection combined with high-speed polygons could be the beginning of multi-chain metrology combined with multi-beam lines using high-power ultra-fast laser radiation.

7.2.1.2 Temporal Scanning
Using a pair of lasers, which are electronically synchronized by asynchronous optical sampling (ASOP), scanning frequencies of the lasers should be achievable up to 10 MHz. A combination with high-velocity polygon mirrors enables an investigation of large areas up to 1 m^2 at very large velocities $v \approx 100$ m/s with large temporal delay ranges $t \approx 1$ ns.

7.2.2
Beam Shaping

7.2.2.1 Spatial Shaping
New optics, using meta materials, have been invented [202–204]. Today these optics are working in the IR spectral range. The attempts to push their applicability into the visible are ongoing and promising first results have been reported about application of meta materials at the $\lambda = 1$ µm wavelength. Meta materials will enable one to image objects using conventional radiation with dimensions well below diffraction limit, also enabling one to focus IR-laser radiation to a spot the size of 10 nm. The nanometer world will be opened for ultra-fast optical metrology by using optical radiation in the UV-VIS-IR spectral range.

7.2.2.2 **Temporal Shaping**

Concepts for temporal beam shaping are under investigation today [77, 205, 206]. The potential of process enhancement by tailored laser radiation is large. Production with flexible product chains need ultra-fast optical metrology to also be flexible: correlated pump and probe methods combined with genetic programs could solve this.

8
Summary

Ultra-fast optical metrology has been presented as a new and innovative tool for engineering technology. The unique properties of optical pump and probe metrology, like the ultra-fast temporal and the ultra-large spatial resolution, are described and new applications as well as new tools for quality inspection, process control and process investigation for applications in ultra-fast engineering technology have been addressed.

The fundamentals of ultra-fast laser sources and their new features compared to conventional laser sources, like the ultrashort pulse duration and the large spectral bandwidth, have been outlined. Laser systems for industrial applications and for ultra-fast optical metrology have been presented, especially the development of high-power laser systems based on fiber technology are becoming industrially relevant.

The temporal and spatial resolution for the ultra-fast optical metrology is achieved by focusing, positioning and scanning the ultra-fast laser radiation. The requirements for the generation of tiny foci as small as 1 μm are described and therefore the properties of Gaussian beams have been highlighted, being the predominant radiation in industrial ultra-fast lasers. The key parameters of ultra-fast laser radiation have been demonstrated for ultra-precise application.

A necessary survey on laser radiation-matter interaction is given, pointing out the ultra-fast laser-induced processes below and above the ablation threshold. Absorption of laser radiation, heating, melting, evaporation and ionization of matter is shortly described and an introduction to plasma physics is given. The ultra-fast physics is elaborated and due to pump and probe metrology, an ultra-fast detection of nearly instantaneous laser-induced processes is enabled.

Important fundamentals on pump and probe techniques have been described, treating the laser radiation as an analyzing beam, discussing the necessary properties, like pulse duration, dispersion and coherence to solve its requirements for pump and probe experiments. The techniques described aim to generate definite states of matter in order to obtain information on processes. The necessary measurement techniques are described, and as a consequence of it the measurement of the probe radiation is derived. Imaging is one task of ultra-fast optical metrology demanding knowledge on diffraction theory. A summary of image formation by microscopy is given for micro- and nano-technology applications using microscope

Ultra-fast Material Metrology. Alexander Horn
Copyright © 2009 WILEY-VCH Verlag GmbH & Co. KGaA, Weinheim
ISBN: 978-3-527-40887-0

objectives. Delay techniques are described, being a central tool for pump and probe metrology. New concepts using double sources are given for ultra-fast scanning.

Ultra-fast detection methods for ultra-fast engineering technology can be subdivided in non-imaging and imaging techniques. In this way ultra-fast laser radiation is adopted to probe the experiment exemplary, due to:

- the large spectral bandwidth necessary for spectroscopy,
- the shapeable temporal pulse adopted for resonant processing,
- the small pulse duration important for the detection of fast processes, and
- the coherent properties used to measure optical parameters in dielectrics.

Different applications of pump and probe metrology for non-imaging and imaging detection have been presented, demonstrating the potential of these methods and examples are presented for selected metals and glasses:

- **Laser drilling of metals** is investigated by detecting the dynamics of the ablated material and the plasma adopting optical pump and probe shadowgraphy.
- **Microstructuring** of metals using ultra-fast laser radiation has been investigated detecting the expansion of the plasma after ablation by using pump and probe speckle photography. The spatial distribution of the shock wave and the ionization front have been imaged. The energy amount coupled into the plasma during ablation has been calculated.
- **Marking of glass** has been investigated detecting the dynamics of crack formation by pump and probe Nomarsky photography. Using transient absorption spectroscopy the electron dynamics in the glass has been monitored.
- **Welding of glass** is feasible due to the properties of ultra-fast laser radiation. The formation of the melt during irradiation has been investigated by pump and probe quantitative phase microscopy. In this way a phase change is time-resolved detected at the front of the welding seam, which makes process control of ultra-fast welding feasible.

There are many perspectives for the future. New laser sources are becoming cheap and reliable, whereas new concepts have been proposed to set up a pump and probe measurement system, like ASOP. New research fields for ultra-fast pump and probe metrology can be generated by the development of new beam shaping techniques to enhance process signals and new optics by enhanced spatial and temporal resolution into the nanometer and attosecond regime.

A societal relevance of optical pump and probe metrology in engineering can be well-demonstrated by looking at the technical developments in semiconductor and micro-technology industry and in the pharmaceutical industry. The undamped down-scaling of feature size taking place needs new engineering technologies, like ultra-fast engineering adopting ultra-fast laser radiation. The pharmaceutical industry increasingly adopts molecular chemistry to generate highly selective drugs, which can be generated by ultra-fast chemistry process controlled by ultra-fast optical metrology.

In this work, the ultra-fast optical pump and probe metrology as a new technique for diagnostics and process control has been integrated successfully into ultra-fast engineering technology.

Appendix A
Lock-in Amplifier

A lock-in amplifier (also known as a phase sensitive detector) is a type of ampli-fier that can extract a signal with a known carrier wave from an extremely noisy environment (signal-to-noise ratio can be as low as –60 dB or even less)[40]. Lock-in amplifiers use mixing, through a frequency mixer, to convert the signal's phase and amplitude to a DC – actually a time-varying low – frequency voltage signal.

In essence, a lock-in amplifier takes the input signal, multiplies it by the refer-ence signal (either provided from the internal oscillator or an external source), and integrates it over a specified time, usually on the order of milliseconds to a few sec-onds. The resulting signal essentially is a DC signal, where the contribution from any signal that is not at the same frequency as the reference signal is attenuated to zero, as well as the out-of-phase component of the signal that has the same fre-quency as the reference signal (because sine functions are orthogonal to the cosine functions of the same frequency), and this is also why a lock-in is a phase sensitive detector.

For a sine reference signal and an input waveform $U_{in}(t)$, the DC output signal $U_{out}(t)$ can be calculated for an analog lock-in amplifier by

$$U_{out}(t) = \frac{1}{T} \int_{t-T}^{t} ds \, \sin\left[2\pi f_{ref} \cdot s + \phi\right] U_{in}(s) \tag{A1}$$

where ϕ is a phase that can be set on the lock-in (set to zero by default).

Practically, many applications of the lock-in technique require just recovering the signal amplitude rather than relative phase to the reference signal. A lock-in amplifier usually measures both in-phase (X) and out-of-phase (Y) components of the signal and can calculate the magnitude (R) from that.

40) http://www.cpm.uncc.edu/programs/
tn1000.pdf

Ultra-fast Material Metrology. Alexander Horn
Copyright © 2009 WILEY-VCH Verlag GmbH & Co. KGaA, Weinheim
ISBN: 978-3-527-40887-0

Appendix B
Onset on Optics

B.1
Abbe Sine Condition

An object in the object plane of an optical system exhibits a transmittance function

$$T(x_0, y_0) = \iint T(k_x, k_y) e^{j(k_{x_0} + k_{y_0})} dk_x dk_y \, . \tag{B1}$$

Assuming no aberrations the image plane coordinates are linearly related to the object plane coordinates by

$$x_i = Mx_0 \, , \tag{B2}$$

$$y_i = My_0 \, , \tag{B3}$$

where M is the system's magnification. Expanding k_{x_0} and k_{y_0} by the magnification M and moving into the image plane one obtains

$$T(x_i, y_i) = \iint T(k_x, k_y) e^{j((k_x/M)x_i + (k_y/M)y_i)} dk_x dk_y \, . \tag{B4}$$

Defining image plane wavenumbers

$$k_x^i = k_x/M \tag{B5}$$

$$k_y^i = k_y/M \tag{B6}$$

the final equation for the image plane field in terms of image plane coordinates and image plane wavenumber is given by

$$T(x_i, y_i) = M^2 \iint T\left(Mk_x^i, Mk_y^i\right) e^{j(k_x^i x_i + k_y^i y_i)} dk_x^i dk_y^i \, . \tag{B7}$$

Ultra-fast Material Metrology. Alexander Horn
Copyright © 2009 WILEY-VCH Verlag GmbH & Co. KGaA, Weinheim
ISBN: 978-3-527-40887-0

The wavenumbers can be expressed in terms of spherical coordinates as

$$k_x = k \sin \theta \cos \phi ,$$
(B8)

$$k_y = k \sin \theta \sin \phi ,$$
(B9)

for which $\phi = 0$, the coordinate transformation between object and image plane wavenumbers takes the form

$$k^i \sin \theta^i = k \sin \theta / M .$$
(B10)

Equation (B10) is the *Abbe sine condition*, reflecting Heisenberg's uncertainty principle for Fourier transform pairs, namely that as the spatial extent of any function is expanded (by the magnification factor, M), the spectral extent contracts by the same factor, M, so that the space-frequency product remains constant.

B.2
Quantitative Phase Microscopy

The software QPm (IATIA) has been adopted using conventional bright-field transmitting or reflected light microscopy without additional optical components [207]. QPm calculates the phase information of an object by using three images of the object taken by a CCD camera at three different object planes of the microscope. This is achieved by moving the object and taking images sequentially.

The algorithm of QPm applied is based on the numerical solution of the intensity transport equation [207–209] and enables the reconstruction of phase information taking conventional bright-field pictures from three different image planes of an investigated object.

The numerical solution results from the solution of the separable Fourier-based wave equation in two planes x and y orthogonal to the propagation direction of light [208, 209]

$$k\frac{\partial I(\vec{r_\perp})}{\partial z} = -\nabla_\perp \cdot \left[I(\vec{r_\perp}) \nabla_\perp \Phi(\vec{r_\perp}) \right]$$

$$\Phi(x, y) = \Phi_x(x, y) + \Phi_y(x, y) ,$$

$$\Phi_x = F^{-1} k_x k_r^{-2} F I^{-1} F^{-1} k_x k_r^{-2} F \left[k\frac{\partial I}{\partial z} \right]$$
(B11)

$$\Phi_y = F^{-1} k_y k_r^{-2} F I^{-1} F^{-1} k_y k_r^{-2} F \left[k\frac{\partial I}{\partial z} \right]$$
(B12)

with $\Phi(x, y)$ representing the reconstructed phase, I the intensity distribution, k the averaged wavenumber $2\pi/\lambda$, λ the wavelength, r the beam propagation coordinate (orthogonal to the optical axis). F describes the Fourier-transformation, F^{-1} the inverse of F (FFT). k_x and k_y are conjugated Fourier variables to the image co-ordinates x and y and $k_r^2 = k_x^2 + k_y^2$. In order to get quantitative values of k_x and k_y

Eqs. (B11) and (B12) have to be rewritten [209] using

$$\Lambda = -\frac{2\pi}{\lambda \Delta z} \frac{1}{(N\Delta x)^2}$$

$$\Phi_x = \Lambda F^{-1} \frac{i}{i^2 + j^2} F \frac{1}{I(x, y)} F^{-1} \frac{i}{i^2 + j^2} F [I_+ - I_-] \tag{B13}$$

$$\Phi_y = \Lambda F^{-1} \frac{j}{i^2 + j^2} F \frac{1}{I(x, y)} F^{-1} \frac{j}{i^2 + j^2} F [I_+ - I_-] \tag{B14}$$

$\Delta k = 1/\Delta x$ represents the Fourier increment with $i, j \in [-N/2, N/2]$. For the FFT a quadratic picture of the dimension $N \times N$ ($N = 2^n$) with the pixel dimension Δx is needed. i and j are the conjugated variables of x and y. In order to consider disturbances like noise or spherical aberrations different filter functions are adopted.

With known displacement Δz, central wavelength λ, intensity information I_0, I_+, I_-, and image scale, the quantitative phase can be calculated. The phase information is strongly related to the spatial resolution: the displacement Δz has to be chosen within the depth of field to get maximal spatial resolution [207]. Utilizing the Köhler illumination is not stringent. Closing the aperture increases the depth of field and the contrast but decreases the applicable objective numerical aperture. The spatial resolution of the detected phase itself depends on the numerical aperture of the objective, the numerical aperture of the condenser, exposure time and CCD gain level. Working with large spatial resolution results in a small resolution of phase. A small displacement Δz results in small differences of the intensity distribution. Resulting from the Shannon-sampling theorem, to reach the maximal spatial resolution each phase point should be detected by 3×3 pixels of the CCD camera. Tables on the dependence of the apposite displacement Δz on the magnification are given in [207]. Noise deteriorates the intensity information and has to be reduced, for example, by cooling the CCD camera, illuminating the CCD camera more than 60% of the saturation intensity, reducing the exposure time and minimizing the gain of the CCD.

After calculation of the quantitative phase different phase microscopy methods, like dark field, differential-interference-, Hoffmann-modulation- and Zernike-phase-contrast can be derived.

Appendix C
Plasma Parameters

C.1
Transport Coefficients

In order to describe the dynamics of a plasma, the characteristic parameters, called transport coefficients, are derived.

C.1.1
Electrical Conductivity

The Drude model is used to estimate the electrical conductivity of a plasma gas and is given by Ohm's law

$$\vec{j}_e = -n_e e \vec{v}_e \sigma_E \vec{E} \tag{C1}$$

with $\vec{v}_e$ the electron velocity, $\vec{j}_e$ the electric current, n_e the electron density and σ_E the electrical conductivity. Assuming that the average time between collisions of electrons with ions is the relaxation time, $\tau = \nu_{ei}$, the velocity

$$\vec{v}_e = \frac{e \vec{E} \tau}{m_e} \tag{C2}$$

gives the outcome of the Newton's law $\dot{\vec{v}}_e = -e \vec{E}/m_e$. Replacing the velocity in Eq. (C1) with (C2) yields the electrical conductivity

$$\sigma_E = \frac{n_e e^2}{m_e \nu_{ei}} . \tag{C3}$$

Equation (C3) yields reliable results for copper $\sigma_E^{Cu} = 5.5 \times 10^{17}$ s^{-1} and, assuming an electron density $n_e \approx 3 \times 10^{-7}$ cm^{-3} of thermal electrons generated by cosmic radiation, one gets the electrical conductivity of air $\sigma_E^{air} \approx 3 \times 10^{-9}$ s^{-1} [91].

C.1.2
Thermal Conductivity

A laser generated plasma never features a spatial uniform temperature distribution. The plasma temperature tends to equilibrate by thermal diffusion. The heat

Ultra-fast Material Metrology. Alexander Horn
Copyright © 2009 WILEY-VCH Verlag GmbH & Co. KGaA, Weinheim
ISBN: 978-3-527-40887-0

flux crossing a unit area per unit time is defined as

$$\vec{q}_H = -\kappa \nabla T \tag{C4}$$

with κ the heat conductivity. κ can be derived by a simple model: the temperature T is a function of x, the particles have an average energy $\varepsilon(x)$, a density n and a velocity v. The flux of the particle through the y-z-plane amounts to $\frac{1}{6}nv$ and the heat flux can be written as

$$q_H = \frac{1}{6}(nv)\left[\varepsilon(x-l) - \varepsilon(x+l)\right] \approx -\frac{1}{3}nvl\frac{\partial\varepsilon}{\partial x} = -\frac{1}{3}nvl\frac{\partial\varepsilon}{\partial T}\frac{\partial T}{\partial x} \tag{C5}$$

with l being the mean free path, Eq. (3.105). The heat capacity c_V at a constant volume is defined as the derivative of energy with respect to the temperature, so the heat flux becomes

$$q_H = -\frac{1}{3}nlc_V\frac{\partial T}{\partial x} \tag{C6}$$

and, compared to Eq. (C4), one obtains, assuming the velocity to be thermal $v = v_T$, for the heat conductivity

$$\kappa = \frac{1}{3}nc_V l v_T = \frac{n_e c_V v_T^2}{3\nu_{ei}} \propto T^{5/2} . \tag{C7}$$

The second equality results using Eq. (3.106). It was assumed that electrons are transporting the heat with $\nu_{ei} \approx T_e^{-3/2}$, $v_T^2 \approx T_e$ and a constant heat capacity c_V.

The Wiedemann–Franz law describes the relation between electrical and thermal conductivity

$$\frac{\kappa}{\sigma_E} = \frac{3}{2}\left(\frac{k_B}{e}\right)^2 T \tag{C8}$$

with the equation of state of an ideal gas $c_V = \frac{3}{2}k_B$.

C.1.3
Diffusivity of Electrons

Electrons in an uniformly distributed plasma with density n_e will begin to move in order to re-equilibrate the plasma after a disturbance in density. These dynamics can be described by the diffusion equation

$$\frac{\partial n_e}{\partial t} = \nabla \cdot (D\nabla n_e) \tag{C9}$$

with D the diffusion coefficient defined by the particle current density

$$\vec{j}_n = n\vec{v} = -D\nabla n . \tag{C10}$$

Similar to the approach given for the thermal conductivity (Section C.1.2) a diffusion coefficient can be derived

$$D = \frac{1}{3}v_T l \propto \frac{T_e^{5/2}}{n_e} \, . \tag{C11}$$

For the derivation of the diffusion coefficient in the presence of a magnetic field in a direction perpendicular to the magnetic field $D_\perp$, one has to generalize Ohm's law

$$\vec{j} = \sigma_E \left(\vec{E} + \frac{\vec{v} \times \vec{B}}{c} \right) \, . \tag{C12}$$

The classical diffusion $D_\perp$ is defined by

$$D_\perp = \frac{c^2 n k_B T}{\sigma_E B^2} \, , \tag{C13}$$

but does not describe experiments well. A semi-empirical formula, called Bohm's diffusion ([91] and citations therein) is given by

$$D_\perp = \frac{c k_B T_e}{16 e B} = D_B \, . \tag{C14}$$

C.1.4
Viscosity

Viscosity arises when adjacent fluid elements flowing with different velocities exchange momentum. The pressure tensor P_{ij} is given by

$$P_{ij} = P\delta_{ij} + \varrho v_i v_j - \eta \left(\frac{\partial v_i}{\partial x_j} + \frac{\partial v_j}{\partial x_i} - \frac{2}{3} \sum_{k=1}^{3} \frac{\partial v_k}{\partial x_k} \delta_{ij} \right) - \zeta \sum_{k=1}^{3} \frac{\partial v_k}{\partial x_k} \delta_{ij} \, , \tag{C15}$$

with the coefficients of viscosity η and ζ used by the Navier–Stokes equation for ordinary fluids

$$\frac{\partial(\varrho v_i)}{\partial t} = -\sum_{k=1}^{3} \frac{\partial P_{ik}}{\partial x_k} \tag{C16}$$

to describe plasma fluidics [210]. In the case of a scalar pressure $P = P(\varrho, T)$, with the plasma density ϱ, and assuming incompressibility like in the case of a fluid, the divergence of the velocity $\nabla \cdot \vec{v} = 0$ vanishes and also the term in Eq. (C15) with the coefficient ζ. In many plasmas the viscosity is of secondary importance. As a first approximation for a plasma flow in the z-direction $\vec{v} = (0, 0, v_z)$, the induced stress P_{xz} is given by

$$P_{xz} = -\left(\frac{1}{3} n m v_T l \right) \frac{\partial v_z}{\partial x} \equiv -\eta \frac{\partial v_z}{\partial x} \tag{C17}$$

as a reasonable scaling law for the mass m, the free path l, the density n and average velocity v_T.

C.2
Debye Length

In a plasma positive and negative particles, for example ions and electrons, tend to screen the Coulomb range of one of the individual particles. The range of the screening is described by the Debye length λ_{De}.

A plasma with an ion charge of Ze at rest surrounded by electrons is neutral with a temperature T_e (the ion temperature is set to $T_i = 0$). The fluid equation of motion for the electrons is given by

$$n_e e \vec{E} + \nabla P_e = 0 \tag{C18}$$

with the electric field $\vec{E}$ and the electronic pressure P_e assumed as an ideal gas $P_e = n_e k_B T_e$. With the electrostatic potential ϕ defined by the electric field $\vec{E} = -\nabla \phi$, the Eq. (C18) results in

$$n_e e \nabla \phi = k_B T_e \nabla n_e \tag{C19}$$

and its solution becomes the electron density distribution for an initial density n_{e0}

$$n_e = n_{e0} \exp\left(\frac{e\phi}{k_B T_e}\right). \tag{C20}$$

The electrostatic potential ϕ results from the Poisson equation

$$\Delta \phi = -4\pi Ze\delta(\vec{r}) + 4\pi e(n_e - n_{e0}). \tag{C21}$$

By inserting Eq. (C20) into the Poisson Eq. (C21) and expanding the exponential function for $e\phi/(k_B T_e) \ll 1$ one obtains the differential equation

$$\left(\Delta - \lambda_{De}^{-2}\right)\phi + 4\pi Ze\delta(\vec{r}) = 0 \tag{C22}$$

with the electron Debye length

$$\lambda_{De} = \left(\frac{k_B T_e}{4\pi e^2 n_{e0}}\right)^{1/2}. \tag{C23}$$

The screening effect by a finite range of the Coulomb interaction is given by the solution of Eq. (C22)

$$\phi = \frac{Ze}{r} \exp\left(-\frac{r}{\lambda_{De}}\right). \tag{C24}$$

The Debye sphere, in which there is a sphere of influence, and outside of which charges are screened, defines the number of electrons N_{De}

$$N_{De} = \frac{4}{3}\pi n_e \lambda_{De}^3, \tag{C25}$$

also called the plasma parameter, within a sphere with radius λ_{De}.

A real plasma has non-zero ion temperatures and therefore the ion Debye shielding has to be considered too. Similar to electrons, the ion density is calculated by Eq. (C20) changing n_e and n_{e0} to n_i and n_{i0}, the charge into Ze and the temperature into T_i

$$n_i = n_{i0} \exp \left(\frac{Ze\phi}{k_B T_i} \right) . \tag{C26}$$

The generalized Poisson equation results in

$$\left(\Delta - \lambda_{De}^{-2} - \lambda_{Di}^{-2} \right) \phi + 4\pi Ze\delta(\vec{r}) = 0 \tag{C27}$$

with the expansion of the exponential function, Eq. (C26), for $Ze\phi/(k_B T_e) \ll 1$ and the ion Debye length

$$\lambda_{Di} = \left(\frac{k_B T_i}{4\pi e^2 n_{i0}} \right)^{1/2} . \tag{C28}$$

The solution of the Poisson Eq. (C27) gives

$$\phi = \frac{Ze}{r} \exp \left(-\frac{r}{\lambda_D} \right) , \tag{C29}$$

with the Debye length

$$\frac{1}{\lambda_D^2} = \frac{1}{\lambda_{De}^2} + \frac{1}{\lambda_{Di}^2} . \tag{C30}$$

C.3
Plasma Oscillations and Waves

Inserted into the equation for charge conservation

$$\nabla \cdot \vec{j}_e + \frac{\partial \varrho_e}{\partial t} = 0 , \tag{C31}$$

the Ohm's law Eq. (C1) and assuming a cold plasma with inert ions, can obtain by taking the derivative in time and using Maxwell's equation $\nabla \cdot \vec{E} = 4\pi \varrho_e$

$$\frac{\partial^2 \varrho_e}{\partial t^2} + \omega_{pe}^2 \varrho_e = 0 \tag{C32}$$

with Eq. (C2) and the plasma frequency

$$\omega_{pe}^2 = \left(\frac{4\pi e^2 n_e}{m_e} \right) . \tag{C33}$$

This differential equation describes the mechanism of plasma oscillation to preserve or restore electrical neutrality, but not to describe wave dynamics. The hydrodynamic equations describe the equation for mass conservation

$$\frac{\partial n_e}{\partial t} + \nabla \cdot (n_e \vec{v}_e) = 0 \tag{C34}$$

and the equation for momentum conservation

$$m_e n_e \left[\frac{\partial \vec{v}_e}{\partial t} + (\vec{v}_e \cdot \nabla)\vec{v}_e \right] = -e n_e \vec{E} - \nabla P_e . \tag{C35}$$

The relation between pressure P and temperature T for a plasma, called equation of state, is given by

$$P_e = n_e k_B T_e . \tag{C36}$$

For an isentropic plasma, in other words at constant entropy like an ideal gas, one can set

$$P_e = C n_e^\gamma , \tag{C37}$$

with the adiabatic exponent γ, defined by the specific heats at constant pressure and volume

$$\gamma = \frac{c_p}{c_V} . \tag{C38}$$

Using Eqs. (C36) and (C37) yields

$$\nabla P_e = P_E \gamma \left(\frac{\nabla n_e}{n_e} \right) = \gamma k_B T_e \nabla n_e . \tag{C39}$$

In combination with the Maxwell equation $\nabla \cdot \vec{E} = 4\pi e(n_i - n_e)$ and the Eqs. (C34), (C35), (C36), and (C39) a set of nonlinear equations is given which cannot be solved analytically. By linearization assuming

$$n_e = n_{e0} + n_{e1} , \quad \vec{v}_e = \vec{v}_{e0} + \vec{v}_{e1} , \quad \vec{E} = \vec{E}_0 + \vec{E}_1 , \tag{C40}$$

with the oscillation amplitudes denoted by the subscript "1" and the equilibrium condition given by

$$n_{e0} = n_i = \text{const.} , \quad \vec{v}_{e0} = 0 , \quad \vec{E}_0 = 0 , \quad \frac{\partial}{\partial t} \left\{ n_{e0}, \vec{v}_{e0}, \vec{E}_0 \right\} = 0 \tag{C41}$$

the mentioned equations are simplified with $(n_{e1}/n_{e0})^2 \ll (n_{e1}/n_{e0})$, $(\vec{v}_{e1} \cdot \nabla)\vec{v}_{e1}) \ll \partial v_{e1}/\partial t$, and so on. To the following set of equations

$$\frac{\partial n_{e1}}{\partial t} + n_{e0} \nabla \cdot \vec{v}_{e1} = 0 ,$$

$$m_e \frac{\partial \vec{v}_{e1}}{\partial t} = -e \vec{E}_1 - \gamma k_B T_e \nabla n_{e1} , \tag{C42}$$

$$\nabla \cdot \vec{E}_1 = -4\pi e n_{e1} .$$

As shown in [91], assuming a one-dimensional problem and using monochromatic waves with a frequency ω with wave number $k = 2\pi/\lambda$ and λ the wavelength, a set of algebraic equations results with the overall solution for the dispersion of the electron plasma wave, called a plasmon, given by

$$\omega^2 = \omega_p^2 + 3k^2 v_{\text{th}}^2 \tag{C43}$$

by using the one-dimensional thermal velocity

$$v_{\text{th}} = \sqrt{\frac{k_B T_e}{m_e}} \ . \tag{C44}$$

By inserting the phase velocity $v_\phi = \omega/k$ and the group velocity $v_g = d\omega/dk$ into the dispersion relation (C43) one gets

$$v_g v_\phi = 6v_{\text{th}}^2 \ . \tag{C45}$$

Plasma waves can propagate by transversal or longitudinal waves, the last representing, for example, the plasmon. In the case of large disturbances of the plasma state, the disturbances grow and the plasma becomes instable, contrary to small disturbances which decay exponentially.

C.4
Coupling Between Electrons and Ions

The critical and ablation surfaces (Figure 3.9) can be strongly coupled to the plasma. The Coulomb interaction for an ideal plasma, like a corona, is weak compared to the thermal energy of the particles. The ratio

$$\Gamma = \frac{E_{Ci}}{E_{Ti}} \quad \Rightarrow \quad \Gamma > 1 \quad \text{strong coupling} \tag{C46}$$

$$\Rightarrow \quad \Gamma < 1 \quad \text{weak coupling by ideal plasma} \tag{C47}$$

defines the coupling parameter with E_{Ci} the Coulomb interaction energy, and E_{Ti} the thermal interaction energy. The sphere each particle occupies is defined by the radius

$$a_k = \left(\frac{3}{4\pi n_k} \right)^{1/3} \quad \text{with} \quad k = e \quad \text{or} \quad i \ . \tag{C48}$$

The strong coupling parameters for ions Γ_{ii}, electrons Γ_{ee} and electron–ions Γ_{ei} are defined by

$$\Gamma_{ii} = \frac{Z^2 e^2}{a_i k_B T_i} \ , \quad \Gamma_{ee} = \frac{e^2}{a_e k_B T_e} \ , \quad \Gamma_{ei} = \frac{Ze^2}{a_e k_B T_e} \ . \tag{C49}$$

In the case of screening by a large number of particles of an ideal plasma, many particles are within the Debye sphere, Eq. (C25). The electron energy distribution can be described by a Maxwell distribution. In plasma with strong coupling the electrons are degenerated and the energy distribution is described by a Fermi–Dirac-distribution.

C.5
Hydrodynamic Instabilities

When high-power laser radiation interacts with a thin foil, plasma is created and accelerates the foil. A Rayleigh–Taylor (RT) instability occurs when a heavy fluid is supported by a light fluid in a gravitational field and also when a light fluid pushes and accelerates a heavy fluid. In the case of the laser-induced plasma, the low-density medium is the plasma, which accelerates the heavy density foil, or in the case of melting and evaporation in coexistence with melt and transient states. These instabilities are found at very large intensities $I > 10^{15}$ W cm^{-2} for example in ICF experiments.

Another class of hydrodynamic instabilities occurring in laser-matter interaction are the Richtmyer–Meshkov and the Kelvin–Helmholtz instabilities:

- the Richtmyer–Meshkov (RM) instabilities occur when a shock wave passes through a corrugated interface between two fluids at different densities and can be considered as the impulsive-acceleration limit of the Rayleigh–Taylor instability. The development of the instability begins with small amplitude perturbations which initially grow linearly with time. This is followed by a nonlinear regime with bubbles appearing in the case of a light fluid penetrating a heavy fluid, and with spikes appearing in the case of a heavy fluid penetrating a light fluid. A chaotic regime is eventually reached and the two fluids mix.
- The Kelvin–Helmholtz (KH) instabilities describe the dynamics of two fluids in contact due to different flow velocities across the interface, for example the generation of water waves by wind blowing over the water surface. The theory can be used to predict the onset of instability and transition to turbulent flow in fluids of different densities moving at various speeds. For short wavelengths, if surface tension can be ignored, two fluids in parallel motion with different velocities and densities will yield an interface that is unstable for all speeds.

The existence of surface tension stabilizes the short wavelength instability, however, and KH instabilities then predict stability until a velocity threshold is reached. For a continuously-varying distribution of density and velocity, (with the lighter layers uppermost, so the fluid is RT-stable), the onset of the KH instability is given by a suitably-defined Richardson number, Ri. The dimensionless Richardson number expresses the ratio of potential to kinetic energy

$$Ri = \frac{gh}{u^2} , \tag{C50}$$

where g is the acceleration due to gravity, h a representative vertical length scale, and u a representative speed. Typically, the layer is unstable for $Ri < 0.25$. Also the study of this instability becomes applicable to inertial confinement fusion (ICF).

Appendix D
Facilities

Different research centers and laser facilities are developing these kind of laser sources, like Lawrence Livermore National Laboratory (LLNL) (Figure 2.7)[41]. This facility uses high-intensity laser radiation for
- laser-plasma interaction under varying plasma conditions,
- Inertial Confinement Fusion (ICF) plasmas,
- Thomson scattering (TS) of stable electron density fluctuations, and
- Compton X-ray generation, collisional X-ray laser based on laser-produced plasma and small capillary discharges.

At the Rutherford Appleton Laboratory's Central Laser Facility in Great Britain, a high-intensity laser, called VULCAN, is used for
- acceleration of electrons for 300 MeV electrons production [211],
- production of the 850th harmonics [212],
- magnetic reconnection in a plasma created by two laser beams [213],
- proton radiography of a laser-driven implosion [214],
- measurements of energy transport patterns in solid density laser plasma interactions at intensities of 5×10^{20} W/cm^2 [215], and
- photon acceleration by laser wake fields [216].[42]

Also at Laboratoire pour l'Utilisation des Lasers Intenses (LULI) at l'Ecole Polytechnique of Paris in France, the interaction of laser radiation with plasmas, the hydrodynamics of laser generated plasmas, the investigation of equation of state at very large pressures, and atomic physics of dense hot plasmas have been investigated by high energy ultra-fast laser radiation.[43]

At the Institute for Laser Engineering (ILE) of the Osaka University, Japan, GEKKO XII is used to investigate the physics of the reactor core plasma including super-high density implosion, plasma heating, and "ultra-high energy density states" on a macroscopic scale [217].[44]

41) http://universitygateway.llnl.gov/institutes/ ilsa/research.html

42) http://www.clf.rl.ac.uk/highlights/HPLS.htm

43) http://www.luli.polytechnique.fr/pages/ themes.htm

44) http://www.ile.osaka-u.ac.jp/zone1/ activities/facilities/spec_e.html#2

Ultra-fast Material Metrology. Alexander Horn
Copyright © 2009 WILEY-VCH Verlag GmbH & Co. KGaA, Weinheim
ISBN: 978-3-527-40887-0

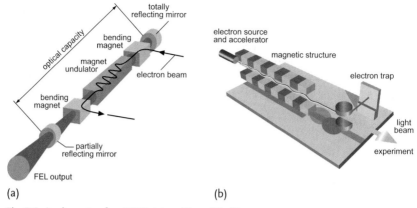

Fig. D.1 A schematic of an IR-FEL (a) and X-ray FEL (b).

Free-electron lasers (FEL) are "exotic laser" systems and in an exceptional position, because the "laser" medium is not a solid or gas but an oscillating free-
electron beam. The amplifying process is called SASE[45]. A FEL emits radiation with
large pulse energies and large repetition rates in a wide spectral range. Different
from solid-state and fiber lasers, FELs are today very large systems occupying much
space, with dimensions of a gym with 1000 m^2. This gives them the widest frequency range of tunable laser type and makes many of them widely tunable, currently
ranging in wavelengths from microwaves, through THz radiation and infrared, to
the visible spectrum, to ultraviolet, to soft X-rays. Around the world actually about
30 FELS are built up and operating experiments[46]. To create a FEL, a beam of electrons (being the seeder) is accelerated to relativistic speeds. The electron beam is
usually generated by the photo effect using UV-ultra-fast laser radiation. The beam
passes through a periodic, transverse magnetic field produced by arranging magnets with alternating poles along the beam path. This array of magnets is called
undulator, or "wiggler", because it forces the electrons to follow a sinusoidal path.
The acceleration of the electrons along this path results in Bremsstrahlung. Since
the photons emitted are related to the electron beam and magnetic field strength,
a FEL can be tuned in frequency. At the Jefferson Lab FEL, for example, a subpicosecond, tunable source from 250 nm to 14 µm with pulse energies up to 300 µJ
at repetition rates up to 75 MHz has been installed. The FEL emits optical radiation
in the infrared with average powers of 10 kW (Figure D.1a)[47]. An ulterior extraordinary FEL with similar set-up is the X-ray FEL. The lack of suitable mirrors in the
extreme ultraviolet and X-ray regimes prevent the operation of an oscillator. Hence,
there must be suitable amplification in a single pass of, for example, the electron
beam through the undulator to make the FEL worthwhile (Figure D.1b)[48]. When
the field extracts enough energy from the electrons over a single pass such that the

45) Self-Amplified Stimulated Emission
46) http://sbfel3.ucsb.edu/www/fel_table.html
47) http://www.srs.dl.ac.uk/Annual_Reports/
AnRep99_00/fel.htm

48) http://www.measurementdb.com/index.php?
name=News&file=article&sid=726

field amplitude cannot be regarded as constant during the FEL process, the FEL operates in the high-gain regime.

THz-, IR-, and UV-FEL have been essential for pioneering physical techniques, including, for example, vibrational photon-echo spectroscopy, material and device physics, nanostructures and nanocrystals, (bio)molecules, fullerenes, clusters and complexes. A more detailed review on FEL application has been given by [218].

Appendix E
List of Abbreviations and Symbols

E.1
Abbreviations

AFM	Atomic Force Microscopy
AOPDF	Acousto-Optic Programmable Dispersive Filter
BPP	Beam Parameter Product
CCD	Charged Coupled Device
CPA	Chirped Pulse Amplification
CPU	Central Processing Unit
DNA	Desoxy Ribonuclein Acid
DRO	Digital Read-Out
EUV	Extreme Ultra-Violet radiation
FEL	Free-Electron Laser
FROG	Frequency-Resolved Optical Gating
GRENOUILLE	Grating-Eliminated No-nonsense Observation of Ultra-fast Incident Laser Light E-fields
GVD	Group Velocity Dispersion
iCCD	intensified CCD
ICF	Inertial Confinement Fusion
ILE	Institute for Laser Engineering
LED	Light Emitting Diode
LLNL	Lawrence Livermore National Laboratory
LTE	Local Thermal Equilibrium
LULI	Laboratoire pour l'Utilisation des Lasers Intenses
MD	Molecular Dynamic calculation
MEMS	Micro-Electro-Mechanical Systems
MOPA	Master Oscillator Power Amplifier
NBOHC	Non-Bridging Oxygen Hole Centers
NIR	Near Infrared Radiation
OPA	Optical Parametric Amplification
OPCPA	Optical Parametric Chirped-Pulse Amplification
PBP	Pulse duration Bandwidth Product
PSF	point spread function

Ultra-fast Material Metrology. Alexander Horn
Copyright © 2009 WILEY-VCH Verlag GmbH & Co. KGaA, Weinheim
ISBN: 978-3-527-40887-0

QEOS	Quotidian Equation Of State
SASE	Self-Amplified Stimulated Emission
SEM	Scanning Electron Microscopy
SHG	Second Harmonic Generation
SNOM	Scanning Nearfield Optical Microscopy
SPIDER	Spectral Phase Interferometry for Direct Electric Field Reconstruction
STE	Self Trapped Excitons
STM	Scanning Tunnel Microscopy
TAS	Transient Absorption Spectroscopy
THz	Teraherz radiation
TOD	Third Order Dispersion
TQPm	Transient quantitative phase microscopy
TS	Thomson scattering
TTM	Two-Temperature Model
ULWD	Ultra-Long Working Distance
VIS	Visible radiation
WLC	White-Light Continuum
X	Roentgen radiation

E.2
Symbols

α	absorption coefficient
α_{eff}	effective optical penetration depth
α_a	avalanche ionization coefficient
b	chirp
$\mathcal{F}'$	focal plane
$\mathcal{V}$	visibility of fringes
χ	dielectric susceptibility tensor
$\chi^{(i)}$	non-linear susceptibility of ith order
δ	skin depth
$\Delta\omega_A$	absorption bandwidth
$\Delta\omega_{\pm}^{\text{SPM}}$	relative broadening of the white light continuum
$\Delta\omega_p$	bandwidth of radiation
$\delta\Phi$	summarized phase change
ΔE	energy difference
Δn	refractive index n is changed
Δn_{Kerr}	refractive index change induced by the Kerr effect
Δt	time difference
Δt	temporal delay
ΔV	volume in the interaction area
Δx	spatial resolution
ε	kinetic energy of the electrons

η	impedance
η	conversion efficiency
$\frac{d\sigma_{ei}}{d\Omega}$	differential scattering cross section
γ	electron–phonon coupling constant
γ	adiabatic exponent
Γ	optical retardation
$\Gamma_{11}(\tau)$	temporal coherence
$\Gamma_{12}(\tau)$	spatial coherence
$\gamma_{12}(\tau)$	complex degree of coherence
γ_e	avalanche ionization coefficient
γ_s	surface tension
$\hbar$	$h/2\pi$
κ	electron heat conductivity
κ	heat conductivity
κ_ν	opacity
κ_{ib}	spatial damping rate of optical energy by inverse Bremsstrahlung
λ	wavelength
Λ	collision corridor
λ_{cc}	cross-correlation wavelength
λ_{De}	electron Debye length
λ_{Di}	ion Debye length
$\mathbf{\Delta}$	speckle pattern translation
$\mathbf{E}$	electric field vector
$\mathbf{P}$	polarization vector
$\mathbf{P}^L$	linear polarization vector
$\mathbf{P}^{NL}$	non-linear polarization vector
$\mathcal{K}$	elliptical integrals of first and second order
μ_0	susceptibility factor
μ_a	absorption coefficient
ν_{ab}	collision frequency
ν_{ei}	electron–ion collision frequency
ω	frequency
$\omega_\pm^{SPM}$	Stokes frequency
ω_{pe}, ω_p	electron plasma frequency
ω_{pi}	ion plasma frequency
ω_l	carrier frequency
ω_L	frequency of laser radiation
ϕ	instantaneous phase of the pulse
Φ	optical phase
Φ	screening effect
ϱ_e	electron charge density
σ	collisional cross-section
σ_{bb}	inter-band (band–band transition) absorption
σ_{coll}	cross-section for collisional absorption by inverse Bremsstrahlung
σ_D	intra-band absorption (inverse Bremsstrahlung)

σ_{ei}	total cross section
σ_{sp}	dimensions of a speckle
σ_x^2, σ_y^2	variances of the intensity distribution
σ_E	electrical conductivity
σ_k	k-photons absorption cross-section
τ	delay time
τ_{abs}	absorption duration
τ_{e-p}	electron–phonon coupling time
τ_c	effective electron–ion collision time
τ_i	lattice heating time
τ_M	characteristic time scale for Marangoni flow
τ_P	characteristic time scale for Pressure-driven flow
θ	divergence angle
θ'	aperture angle
θ^{real}	divergence of technical radiation
θ_0	entrance angle
θ_s	beam stability
$\tilde{\mathcal{E}}(\eta, \xi)$	electrical field in the retarded coordinates
$\tilde{E}$	complex amplitude of the electrical Field
$\tilde{P}$	complex amplitude of the polarization
$\tilde{P}^L$	linear polarization in the frequency space
ε	dielectric constant
φ	phase
$\vec{j}_e$	electrical conductivity
$\vec{j}_n$	current density
$\vec{f}_p$	ponderomotive force
$\vec{q}_H$	heat flux
ξ, η	retarded spatial and temporal coordinates
ζ	Guy phase
ζ	plasma vacuum interface
a	chirp parameter
A	absorption coefficient, absorptivity
A_j	ion species
b	confocal parameter
b	impact parameter
c	speed of light
C_e	electron heat capacities at constant volume
C_i	ion heat capacities at constant volume
c_T	speed of sound
D	thermal diffusivity
D	diffusion coefficient
dF	collision probability
E	electrical field
f	focal length
F	fluence

f	sweeping rate
F	free energy
F_{thr}	threshold fluence for ablation
F_a	particle flux
f_p	repetition rate
I	intensity
I_j	self-coherence
J_{12}	mutual coherence
j_{ie}	induced emission
j_{se}	spontaneous emission
j_a	absorption and scattering
J_e	current
k	wave number
k_B	Boltzmann constant
k_l	carrier wave number
k_l''	group velocity dispersion
L	moment of inertia
L	thickness of the optical element
l_{ca}	closest approach
l_a	mean free path
L_c	coherence length
M	propagation matrix
m	mass
M^2	beam quality
m_{eff}	effective electron mass
M_T	matrix for reflection at concave mirror
N	multi photon factor
n	refractive index
n_{ec}	critical electron density
n_2	second-order nonlinear refractive index
N_e	electron number
n_e	electron density
$N_e(\varepsilon, t)$	electron density distribution
n_R	refractive index of plasma
NA	numerical Aperture
P_{c1}	threshold for self-focusing
P_{ij}	pressure tensor
P_a	recoil pressure
P_C	electron pressure generated by the "cold electrons"
P_H	electron pressure generated by the "hot electrons"
P_i	pressure generated by ions
P_L	radiation pressure
Q	Joule heating
Q_j	partial function
R	wave front radius

r	shock front radius
$R(\Omega)$	response function of the WLC-interaction region
$r(T_h)$	ratio of the energies S_1 and S_2
r_0	radius of the lens aperture
R_1, R_2	radii of curvature of the lens surfaces
Ri	Richardson number
S	sources and drains of electrons
s_x, s_y	space frequency components
T	temperature
T	transmittance
t_{event}	duration of the event
t_{WLC}^{cc}	pulse duration of chirped WLC
T_e	electron temperature
T_e^C	cold electron temperature
T_e^H	hot electron temperature
T_i	ion temperature
t_p	pulse duration
U	energy transfer rate from electrons to ions
$U(P)$	electric field in P
v	velocity
$v(t)$	scan velocity
v_{eff}	effective electron velocity
v_{Te}	thermal electron energy
v_E	electron velocity
V_F	focal volume
v_g	group velocity
v_p	phase velocity
w	beam radius
$w(E, T_h)$	spectral energy density
w, Y	resolving power
w_{PD}	energy of the X-rays
$W_{x,0}$	focal beam radius of technical radiation
w_0	focus beam radius
w_L	beam radius in front of a lens
W_x	beam radius of technical radiation
$Z_{R,x}$	Rayleigh length of technical radiation
z_R	Rayleigh length

References

1 Hänsch, T.W. (2005) Nobel lecture: Passion for precision. *Reviews of Modern Physics*, **78**, 1297–1309.

2 Weck, A., Crawford, T.H.R., Wilkinson, D.S., Haugen, H.K., and Preston, J.S. (2007) Laser drilling of high aspect ratio holes in copper with femtosecond, picosecond and nanosecond pulses. *Applied Physics A*, **90**(3), 537–543.

3 Fermann, M.E., Galvanauskas, A., and Sucha, G. (2003) *Ultrafast Lasers: Technology and Applications*, Marcel Decker Inc., New York.

4 Wheatstone, C (1879) *The Scientific Papers of Sir Charles Wheatstone*, Published by the Physical Society of London, Taylor and Francis.

5 Krehl, P. and Engemann, S. (1995) August Toepler – The first who visualized shock waves. *Shock Waves*, **5**(1), 1–18.

6 Lawrence, E.O. and Dunnington, F.G. (1930) On the early stages of electric sparks. *Physical Review*, **35**(4), 396–407.

7 Freed, S. (1934) On the velocity of light. *Physical Review*, **46**(11), 1025.

8 Edgerton, H.E. (1979) *Electronic Flash, Strobe*, MIT Press.

9 Edgerton, H.E. and Killian, J.R. (1939) *Flash!: Seeing the Unseen by Ultra Highspeed Photography*, Hale, Cushman & Flint.

10 Edgerton, H.E. (1949) Electric system, including a vaporelectric discharge device, US Patent 2,478,902, 16 August 1949.

11 Elkins, J. (2004) Harold Edgerton's rapatronic photographs of atomic tests. *History of photography*, **28**(1), 74–81.

12 Moynihan, M.F. (2000) The scientific community and intelligence collection. *Physics Today*, **53**(12), 51–56.

13 Davidhazy, A. (1963) Making a streak camera. *Photo Methods for Industry (PMI) magazine*.

14 Wolfman, A. *PMI, Photo Methods for Industry*, Gellert Pub. Corp, 1958–1974.

15 Maiman, T.H. (1960) Stimulated optical radiation in ruby. *Nature*, **187**, 493–494.

16 Schäfer, F.P. and Drexhage, K.H. (1990) *Dye Lasers*, Springer Verlag.

17 Schmidt, W. and Schäfer, F.P. (1968) Self-mode-locking of dye-lasers with saturated absorbers. *Physics Letters A*, **26**(11), 558–559.

18 Shank, C.V., Ippen, E.P., and Bersohn, R. (1976) Time-resolved spectroscopy of hemoglobin and its complexes with subpicosecond optical pulses. *Science*, **193**(4247), 50.

19 Bartana, A., Korsloff, R., and Tannor, D.J. (1993) Laser cooling of molecular internal degrees of freedom by a series of shaped pulses. *The Journal of Chemical Physics*, **99**(1), 196–210.

20 Tannor, D. and Bartana, A. (1999) On the interplay of control fields and spontaneous emission in laser cooling. *Journal of Physical Chemistry A*, **103**(10359), 165.

21 Zewail, A.H. (2000) Femtochemistry: Atomic-scale dynamics of the chemical bond. *Journal of Physical Chemistry A*, **104**, 5660–5694.

22 Kurz, H. and Bloembergen, A.N. (eds) (1985) *Picosecond Photon-Solid interaction*, Vol. 35. Materials Research Society Symposium Proceedings

Ultra-fast Material Metrology. Alexander Horn
Copyright © 2009 WILEY-VCH Verlag GmbH & Co. KGaA, Weinheim
ISBN: 978-3-527-40887-0

23 Hentschel, M., Kienberger, R., Spielmann, C., Reider, G.A., Milosevic, N., Brabec, T., Corkum, P., Heinzmann, U., Drescher, M., and Krausz, F. (2001) Attosecond metrology. *Nature*, **414**, 509–513.

24 Drescher, M., Hentschel, M., Kienberger, R., Uiberacker, M., Yakovlev, V., Scrinzi, A., Westerwalbesloh, T., Kleineberg, U., Heinzmann, U., and Krausz, F. (2002) Time-resolved atomic inner-shell spectroscopy. *Nature*, **419**, 803–807.

25 Baltuska, A., Udem, T., Uiberacker, M., Hentschel, M., Goulielmakis, E., Gohle, Yakovlev, V.S., Scrinzi, A., Hansch, T.W., and Krausz, F. (2003) Attosecond control of electronic processes by intense light fields. *Nature*, **421**(6923), 611–615.

26 Schaller, R.R. (1997) Moore's law: past, present and future. *Spectrum, IEEE*, **34**(6), 52–59.

27 Knoesel, E., Hotzel, A., and Wolf, M. (1998) Ultrafast dynamics of hot electrons and holes in copper: Excitation, energy relaxation, and transport effects. *Physical Review B*, **57**(20), 12812–12824.

28 Pawlik, S., Bauer, M., and Aeschlimann, M. (1997) Lifetime difference of photoexcited electrons between intraband and interband transitions. *Surface Science*, **377**, 379.

29 White, J.O., Cuzeau, S., Hulin, D., and Vanderhaghen, R. (1998) Subpicosecond hot carrier cooling in amorphous silicon. *Journal of Applied Physics*, **84**(9), 4984–4991.

30 Callan, J.P., Kim, A.M.T., Huang, L., and Mazur, E. (2000) Ultrafast electron and lattice dynamics in semiconductors at high excited carrier densities. *Chemical Physics*, **251**(1–3), 167–179.

31 Breitling, D., Ruf, A., Berger, P.W., Dausinger, F.H., Klimentov, S.M., Pivovarov, P.A., Kononenko, T.V., and Konov, V.I. (2003) Plasma effects during ablation and drilling using pulsed solid-state lasers. *Proceedings of the SPIE*, **5121**, 24.

32 Breitling, D., Müller, K.P., Ruf, A., Berger, P., and Dausinger, F. (2003) Material-vapor dynamics during ablation with ultrashort pulses. *Fourth International Symposium on Laser Precision Microfabrication*, (eds I. Miyamoto, A. Ostendorf, K. Sugioka, H. Helvajian). Proceedings of the SPIE, **5063**, 81–86.

33 Temnov, V.V., Sokolowski-Tinten, K., Zhou, P., and von der Linde, D. (2004) Femtosecond time-resolved interferometric microscopy. *Applied Physics A: Materials Science & Processing*, **78**(4), 483–489.

34 Rulliere, C. (1998) *Femtosecond Laser Pulses*, Springer Verlag, Berlin, Heidelberg, New York.

35 Russbüldt, P. (2005) Design und Analyse kompakter, Diodengepumter Femtosekunden-laser. Ph.D. thesis. RWTH-Aachen.

36 Cahill, D.G. and Yalisove, S.M. (2006) Ultrafast lasers in materials research. *MRS Bulletin*, **31**, 594–600.

37 Backus, S., Durfee III, C.G., Murnane, M.M., and Kapteyn, H.C. (1998) High power ultrafast lasers. *Review of Scientific Instruments*, **69**, 1207.

38 Schibli, T.R., Kuzucu, O., Kim, J.W., Ippen, E.P., Fujimoto, J.G., Kaertner, F.X., Scheuer, V., and Angelow, G. (2003) Toward single-cycle laser systems. *IEEE Journal of Selected Topics in Quantum Electronics*, **9**(4), 990–1001.

39 Sutter, D.H., Steinmeyer, G., Gallmann, L., Matuschek, N., Morier-Genoud, F., Keller, U., Scheuer, V., Angelow, G., and Tschudi, T. (1999) Semiconductor saturable-absorber mirror assisted Kerr-lens mode-locked Ti:sapphire laser producing pulses in the two-cycle regime. *Optics Letters*, **24**(9), 631–633.

40 Udem, T., Reichert, J., Holzwarth, R., and Hänsch, T.W. (1999) Absolute optical frequency measurement of the cesium D_1 line with a mode-locked laser. *Physical Review Letters*, **82**(18), 3568–3571.

41 Apolonski, A., Poppe, A., Tempea, G., Spielmann, C., Udem, T., Holzwarth, R., TW Haensch, and Krausz, F. (2000) Experimental access to the absolute phase of few-cycle light pulses. *Physical Review Letters*, **85**, 740–743.

42 Shelton, R.K., Ma, L.-S., Hall, J.L., Kapteyn, C., and Murnaneand, M.M., Jun, Y. (2001) Coherent pulse synthesis from two (formerly) independent passively mode-locked Ti:sapphire oscillators. *Lasers and Electro-Optics*, 2001, CLEO. Technical Digest. Summaries of papers presented at the Conference, pp. CPD10–CP1-2.

43 Yamada, E., Takara, H., Ohara, T., Sato, K., Morioka, T., Jinguji, K., Itoh, M., and Ishii, M. (2001) A high SNR, 150 ch supercontinuum CW optical source with precise 25 GHz spacing for 10 Gbit/s DWDM systems. *Optical Fiber Communication Conference and Exhibit*, 2001, OFC 2001, 1.

44 Takara, H., Yamada, E., Ohara, T., Sato, K., Jinguji, K., Inoue, Y., Shibata, T., and Morioka, T. (2001) 106 × 10 Gbit/s, 25 GHz-spaced, 640 km DWDM transmissionemploying a single supercontinuum multi-carrier source. *Lasers and Electro-Optics*, 2001. CLEO'01. Technical Digest. CPD11–CPI-2

45 Mark, J., Liu, L.Y., Hall, K.L., Haus, H.A., and Ippen, E.P. (1989) Femtosecond pulse generation in a laser with a nonlinear external resonator. *Optics Letters*, 14(1), 48–50.

46 Ippen, E.P., Haus, H.A., and Liu, L.Y. (1989) Additive pulse mode locking. *Journal of the Optical Society of America B*, 6(9), 1736–1745.

47 Strickland, D. and Mourou, G. (1985) Compression of amplified chirped optical pulses. *Optics Communications*, 56, 219.

48 Sprangle, P., Esarey, E., Ting, A., and Joyce, G. (1988) Laser wakefield acceleration and relativistic optical guiding. *Applied Physics Letters*, 53, 2146.

49 Rousse, A., Phuoc, K.T., Shah, R., Pukhov, A., Lefebvre, E., Malka, V., Kiselev, S., Burgy, F., Rousseau, J.P., Umstadter, D., and Hulin, D. (2005) Production of a keV X-ray beam from synchrotron radiation in relativistic laserplasma interaction. *Physical Review Letters*, 93, 135005.

50 Mourou, G. and Umstadter, D. (2002) Extreme light. *Scientific American Digital*, 286, 80, The Edge of Physics, Special Editions.

51 Plaessmann, H. and Grossman, W.M. (1997) Multi-pass light amplifier, US Patent 5,615,043, 25 March 1997.

52 Jesse, K. (2005) *Femtosekundenlaser*, Springer, Berlin, Heidelberg, New York.

53 Shah, L., Fermann, M.E., Dawson, J.W., and Barty, C.P.J. (2006) Micromachining with a 50 W, 50 μJ, subpicosecond fiber laser system. *Optics Express*, 14, 12546–12551.

54 Schnitzler, C., Hofer, M., Luttmann, J., Hoffmann, D., Poprawe, R. (2002) A cw kW-class diode end pumped Nd:YAG slab laser. In: Lasers and Electro-Optics, 2002. CLEO'02. Technical Digest, pp. 766–768.

55 Du, K., Wu, N., Xu, J., Giesekus, J., Loosen, P., and Poprawe, R. (1998) Partially end-pumped Nd:YAG slab laser with a hybrid resonator. *Optics Letters*, 23(5), 370–372.

56 Russbüldt, P., Hoffmann, H.D., and Mans, T.G. (2009) Power scaling of ytterbium INNOSLAB amplifiers beyond 100 W average power. *Proceedings of the SPIE, Photonics West 2009*, 23, 7193.

57 Russbüldt, P., Rotarius, G.A.P., Mans, T.G., Hoffmann, H.D., Poprawe, R., Eidam, T., Limpert, J., and Tünnermann, A. (2009) Hybrid 400 W Fiber-Innoslab fs-Amplifier. *Advanced Solid-State Photonics, OSA Tech. Digest (Postdeadline)*.

58 Poprawe, R., Gillner, A., Hoffmann, D., Gottmann, J., Wawers, W., Schulz, W., Phipps, C.R. (2008) High-power laser ablation VII: 20–24 April 2008, Taos, New Mexico, USA. Bellingham, WA: SPIE, 2008. (SPIE Proceedings Series 7005), Paper 700502.

59 Lindenberg, A.M., Larsson, J., Sokolowski-Tinten, K., Gaffney, K.J., Blome, C., Synnergren, O., Sheppard, J., Caleman, C., MacPhee, A.G., and Weinstein, D. (2005) Atomic-scale visualization of inertial dynamics. *Science (Washington)*, 308(5720), 392–395.

60 Corkum, P.B. and Krausz, F. (2007) Attosecond science. *Nature Physics*, 3(6), 381–387.

61 Alda, J. (2003) Laser and gaussian beam propagation and transformation. *Encyclopedia of Optical Engineering*, pp. 999–1013, doi:10.1081/E-EOE 120009751.

62 Born, M. and Wolf, E. (1980) *Principles of Optics*, Pergamon Press, New York.

63 Siegman, A.E. (1994) Defining and measuring laser beam quality. *Solid State Lasers: New Developments and Applications*, pp. 13–28.

64 Bor, Z. (1989) Distortion of femtosecond laser pulses in lenses. *Optics Letters*, **14**(2), 119–121.

65 Zhu, G., van Howe, J., Durst, M., Zipfel, W., and Xu, C. (2005) Simultaneous spatial and temporal focusing of femtosecond pulses. *Optics Express*, **13**(6), 2153–2159.

66 Guha, S. and Gillen, G.D. (2007) Vector diffraction theory of refraction of light by a spherical surface. *Journal of the Optical Society of America B*, **24**(1), 1–8.

67 Dhayalan, V., Standnes, T., Stamnes, J.J., and Heier, H. (1997) Scalar and electromagnetic diffraction point-spread functions for high-NA microlenses. *Pure Applied Optics*, **6**, 603–615.

68 Siegman, A.E. (1986) *Lasers*, University Science Books, Mill Valley.

69 Wright, D., Greve, P., Fleischer, J., and Austin, L. (1992) Laser beam width, divergence and beam propagation factor an international standardization approach. *Optical and Quantum Electronics*, **24**(9), 993–1000.

70 Siegman, A.E. (1990) Laser beam propagation and beam quality formulas using spatial-frequency and intensity-moments analyses. *Draft Version*, **49**(2), 1–22.

71 Siegman, A.E. (1990) New developments in laser resonators. *SPIE Optical Resonators*, **1224**, 2–8.

72 Diels, J.-C. and Rudolph, W. (1996) *Ultrashort Laser Pulse Phenomena*, Academic Press, San Diego.

73 Bor, Z. (1988) Distortion of Femtosecond Laser Pulses in Lenses and Lens Systems. *Journal of Modern Optics*, **35**(12), 1907–1918.

74 Korte, F., Adams, S., Egbert, A., Fallnich, C., Ostendorf, A., Nolte, S., Will, M., Ruske, J.P., Chichkov, B., and Tuennermann, A. (2000) Sub-diffraction limited structuring of solid targets with femtosecond laser pulses. *Optics Express*, **7**(2), 41–49.

75 Stoian, R., Boyle, M., Thoss, A., Rosenfeld, A., Ashkenasi, D., Korn, G., Campbell, E.E.B., and Hertel, I.V. (2002) Ultrafast laser ablation of dielectrics employing temporally shaped femtosecond pulses. *Proceedings of the SPIE*, **4426**, 78–81.

76 Stoian, R., Boyle, M., Thoss, A., Rosenfeld, A., Korn, G., and Hertel, I.V. (2003) Dynamic temporal pulse shaping in advanced ultrafast laser material processing. *Applied Physics A: Materials Science & Processing*, **77**(2), 265–269.

77 Englert, L., Rethfeld, B., Haag, L., Wollenhaupt, M., Sarpe-Tudoran, C., and Baumert, T. (2007) Control of ionization processes in high band gap materials via tailored femtosecond pulses. *Optics Express*, **15**(26), 17855–17862.

78 Korte, F., Nolte, S., Chichkov, B.N., Bauer, T., Kamlage, G., Wagner, T., Fallnich, C., and Welling, H. (1999) Far-field and near-field material processing with femtosecond laser pulses. *Applied Physics A: Materials Science & Processing*, **69**(7), 7–11.

79 Georg, R.A. (2000) *Linearlager und Linearführungssysteme*, expert Verlag, Renningen-Malmsheim.

80 Chung, S.H., Clark, D.A., Gabel, C.V., Mazur, E., and Samuel, A.D.T. (2006) The role of the AFD neuron in C. elegans thermotaxis analyzed using femtosecond laser ablation. *BMC Neuroscience*, **7**(1), 30.

81 Koch, J., Korte, F., Bauer, T., Fallnich, C., Ostendorf, A., and Chichkov, B.N. (2005) Nanotexturing of gold films by femtosecond laser-induced melt dynamics. *Applied Physics A: Materials Science & Processing*, **81**(2), 325–328.

82 Menzel, R. (2001) *Photonics*, Springer Verlag, Berlin, Heidelberg.

83 Marburger, J.H. (1975) Self-focusing: Theory. *Progress in Quantum Electronics*, **4**, 35–110.

84 Yang, G.Y. and Shen, Y.R. (1984) Spectral broadening of ultrashort pulses in a nonlinear medium. *Optics Letters*, **9**(11), 510–512.

85 Brodeur, A. and Chin, S.L. (1999) Ultrafast white-light continuum generation and self-focussing in transparent condesed media. *Journal of the Optic Society of America B*, **16**(4), 637–650.

86 Luther, G.G., Newell, A.C., Moloney, J.V., and Wright, E.M. (1994) Short-pulse conical emission and spectral broadening in normally dispersive media. *Optics Letters*, **19**(11), 789–791.

87 Kosareva, O.G., Kandidov, V.P., Brodeur, A., Chien, C.Y., and Chin, S.L. (1997) Conical emission from laser-plasma interactions in the filamentation of powerful ultrashort laser pulses in air. *Optics Letters*, **22**(17), 1332–1334.

88 Herrmann, R.F.W., Gerlach, J., and Campbell, E.E.B. (1998) Ultrashort pulse laser ablation of silicon: an MD simulation study. *Applied Physics A: Materials Science & Processing*, **66**(1), 35–42.

89 Nedialkov, N.N., Imamova, S.E., and Atanasov, P. (2004) Ablation of metals by ultrashort laser pulses. *Journal of Physics D, Applied Physics*, **37**(4), 638–643.

90 Zhigilei, L.V. (2003) Dynamics of the plume formation and parameters of the ejected clusters in short-pulse laser ablation. *Applied Physics A: Materials Science & Processing*, **76**(3), 339–350.

91 Eliezer, S. (2002) *The Interaction of High Power Lasers with Plasmas*, Institute of Physics Publishing.

92 Fisher, A.J., Hayes, W., and Stoneham, A.M. (1990) Theory of the structure of the self-strapped exciton in quartz. *Journal of Physics: Condensed Matter*, **2**, 6707–6720.

93 Sun, C.-K., Vallée, F., Acioli, L.H., Ippen, E.P., and Fujimoto, J.G. (1994) Femtosecond-tunable measurement of electron thermalization in gold. *Physical Review B*, 50(20), 15337–15348.

94 Fisher, D., Fraenkel, M., Henis, Z., Moshe, E., and Eliezer, S. (2001) Interband and intraband (Drude) contributions to femtosecond laser absorption in aluminum. *Physical Review E*, 65(1), 16409.

95 Perry, M.D., Stuart, B.C., Banks, P.S., Feit, M.D., Yanocsky, V., and Rubenchik, A.M. (1999) Ultrashort-pulse laser machining of dielectric materials. *Journal of Applied Physics*, **85**(9), 6803–6810.

96 Anisimov, S.I., Kapeliovich, B.L., and Perel'Man, T.L. (1974) Electron emission from metal surfaces exposed to ultrashort laser pulses. *Soviet Physics – JETP*, **39**, 375.

97 Ginzburg, V.L. (1970) The propagation of electromagnetic waves in plasmas. *International Series of Monographs in Electromagnetic Waves*, Pergamon, Oxford, 2nd rev. and enl. ed.

98 Russo, R.E., Mao, X., Gonzalez, J.J., and Mao, S.S. (2002) Femtosecond laser ablation ICP-MS. *Journal of Analytical Atomic Spectrometry*, **17**(9), 1072–1075.

99 Herrmann, R.F.W., Gerlach, J., and Campbell, E.E.B. (1998) Ultrashort pulse laser ablation of silicon: an md simulation study. *Applied Physics A*, **66**, 35–42.

100 Prokhorov, A.M. *et al.* (1990) *Laser Heating of Metals*, Hilger.

101 Preuss, S., Späth, M., Zhang, Y., and Stuke, M. (1993) Time resolved dynamics of subpicosecond laser ablation. *Applied Physics Letters*, **62**(23), 3049–3051.

102 Krüger, J. and Kautek, W. (eds) (1996) *Sub-picosecond-pulse laser machining of advanced materials*, Vol 2. ECLAT'96.

103 Chichkov, B.N., Momma, C., Nolte, S., von Alvensleben, F., and Tünnermann, A. (1996) Femtosecond, picosecond and nanosecond laser ablation of solids. *Applied Physics A*, **63**, 109–115.

104 Ben-Yakar, A., Harkin, A., Ashmore, J., Byer, R.L., and Stone, H.A. (2007) Thermal and fluid processes of a thin melt zone during femtosecond laser ablation of glass: the formation of rims by single laser pulses. *Journal of Physics D, Applied Physics (Print)*, **40**(5), 1447–1459.

105 Ben-Yakar, A. (2004) Femtosecond laser ablation properties of borosilicate glass. *Journal of Applied Physics*, **96**(9), 5316.

106 Choi, T.Y. and Grigoropoulos, C.P. (2002) Plasma and ablation dynamics in ultrafast laser processing of crystalline silicon. *Journal of Applied Physics*, **92**, 4918.

107 Ye, M. and Grigoropoulos, C.P. (2001) Time-of-flight and emission spectroscopy study of femtosecond laser ablation of titanium. *Journal of Applied Physics*, **89**, 5183.

108 Dausinger, F., Lichtner, F., and Lubatschowski, H. (2004) *Femtosecond Technology for Technical and Medical Applications*, Springer.

109 Stoian, R., Rosenfeld, A., Ashkenasi, D., Hertel, I.V., Bulgakova, N.M., and Campbell, E.E.B. (2002) Surface charging and impulsive ion ejection during ultrashort pulsed laser ablation. *Physical Review Letters*, **88**(9), 97603.

110 Perez, D. and Lewis, L.J. (2002) Ablation of solids under femtosecond laser pulses. *Physical Review Letters*, **89**(25), 255504.

111 Sokolowski-Tinten, K., Bialkowski, J., Cavalleri, A., Boing, M., Schueler, H., and D. von der Linde. (2003) Dynamics of femtosecond-laser-induced ablation from solid surfaces. *Proceedings of the SPIE*, **3343**, 46.

112 Sokolowski-Tinten, K., Bialkomiski, J., Boing, M., Cavalleri, A., and von der Linde, D. (1999) Bulk phase explosion and surface boiling during short pulse laserablation of semiconductors. *Quantum Electronics and Laser Science Conference, 1999. Technical Digest. Summaries of Papers Presented at the*, pp. 231–232.

113 Ruf, A. (2004) *Modellierung des Perkussionsbohrens von Metallen mit kurz- und ultrakurzgepulsten Lasern*, Utz.

114 Breitling, D., Ruf, A., and Dausinger, F. (2004) Fundamental aspects in machining of metals with short and ultrashort laser pulses. *Proceedings of the SPIE*, **5339**, 49–63.

115 Ladieu, F., Martin, P., and Guizard, S. (2002) Measuring thermal effects in femtosecond laser-induced breakdown of dielectrics. *Applied Physics Letters*, **81**(6), 957–959.

116 Koubassov, V., Laprise, J.F., Théberge, F., Förster, E., Sauerbrey, R., Müller, B., Glatzel, U., and Chin, S.L. (2004) Ultrafast laser-induced melting of glass. *Applied Physics A: Materials Science & Processing*, **79**(3), 499–505.

117 Nolte, S., Momma, C., Jacobs, H., Tünnermann, A., Chichkov, B.N., Wellegehausen, B., and Welling, H. (1997) Ablation of metals by ultrashort laser pulses. *Journal of the Optical Society of America B*, **14**(10), 2716–2722.

118 Tien, A.-C., Backus, S., Kapteyn, H., Murnane, M., and Mourou, G. (1999) Short-pulse laser damage in transparent materials as a function of pulse duration. *Physical Review Letters*, **82**(19), 3883–3886.

119 Hippel, A. (1932) Elektrische Festigkeit und Kristallbau. *Zeitschrift für Physik A Hadrons and Nuclei*, **75**(3), 145–170.

120 Yablonovitch, E. and Bloembergen, N. (1972) Avalanche ionization and the limiting diameter of filaments induced by light pulses in transparent media. *Physical Review Letters*, **29**(14), 907–910.

121 Bloembergen, N. (1974) Laser-induced electric breakdown in solids. *IEEE Jounal of Quantum Electronics*, QE-**10**(3), 375–386.

122 Kaiser, A., Rethfeld, B., Vicanek, M., and Simon, G. (2000) Microsopic processes in dielectrics under irradiation by subpicosecond laser pulses. *Physical Review B*, **61**(17), 11437–11450.

123 Rethfeld, B., Kaiser, A., Vicanek, M., and Simon, G. (2000) Microscopical dynamics in solids absorbing a subpicosecond laser pulse. *Proceedings of the SPIE*, **4065**, 356–370.

124 Stuart, B.C., Feit, M.D., Rubenchik, A.M., Shore, B.W., and Perry, M.D. (1995) Laser-induced damage in dielectrics with nanosecond to subpicosecond pulses. *Physical Review Letters*, **74**(12), 2248–2251.

125 Gamaly, E.G., Rode, A.V., Tikhonchuk, V.T., and Luther-Davies, B. (2002) Electrostatic mechanism of ablation by femtosecond lasers. *Applied Surface Science*, **197–198**, 699–704.

126 Keldish, L.V. (1965) Ionization in the field of a strong electromagnetic wave. *Soviet Physics – JETP*, **20**(5), 1307–1314.

127 Schmidt, G. (1979) *Physics of High Temperature Plasmas*, Academic Press, Inc., New York, p. 420.

128 Landau, L.D. and Lifshitz, E.M. (1980) *The Classical Theory of Fields*, Butterworth-Heinemann.

129 Kupersztych, J. (1985) Electron acceleration in high-frequency longitudinal waves, Doppler-shifted ponderomotive forces, and Landau damping. *Physical Review Letters*, **54**(13), 1385–1387.

130 James, D., Savedoff, M., and Wolf, E. (1990) Shifts of spectral lines caused by scattering from fluctuating random media (quasar spectrum analysis). *Astrophysical Journal*, **359**, 67–71.

131 Trebino, R. (2002) *Frequency-Resolved Optical Gating: The Measurement of Ultrashort Laser Pulses*, Kluwer Academic Publishers.

132 Hecht, E. (1998) *Optics*, Addison-Wesley, New York.

133 Sucha, G., Fermann, M.E., Harter, D.J., and Hofer, M. (1996) A new method for rapid temporal scanning of ultrafast lasers. *Selected Topics in Quantum Electronics, Journal of IEEE*, **2**(3), 605–621.

134 Mingareev, I., Horn, A., and Kreutz, E.W. (2006) Observation of melt ejection in metals up to 1 µs after femtosecond laser irradiation by a novel pump-probe photography setup. *Proceedings of SPIE*, **6261**, 62610A.

135 Mingareev, I. (2009) Ultrafast dynamics of melting and ablation at large laser intensities. PhD thesis. RWTH-Aachen.

136 Siegman, A.E. (1974) An Introduction to Lasers and Masers. *American Journal of Physics*, **42**(6), 521–529.

137 Herriott, D., Kogelnik, H., and Kompfner, R. (1964) Off-axis paths in spherical mirror interferometers. *Applied Optics*, **3**(4), 523–526.

138 Piyaket, R., Hunter, S., Ford, J.E., and Esener, S. (1995) Programmable ultrashort optical pulse delay using an acousto-optic deflector. *Applied Optics*, **34**(8), 1445–1453.

139 Edelstein, D.C., Romney, R.B., and Scheuermann, M. (1991) Rapid programmable 300 ps optical delay scanner and signal-averaging system for ultrafast measurements. *Review of Scientific Instruments*, **62**, 579.

140 Horn, A., Mingareev, Gottmann, J., Werth, A., and Brenk, U. (2007) Dynamical detection of optical phase changes during micro-welding of glass with ultrashort laser radiation. *Measurement Science and Technology*, **18**, 1–6.

141 Horn, A., Khajehnouri, H., Kreutz, E.W., and Poprawe, R. (2002) Ultrafast pump and probe investigations on the interaction of femtosecond laser pulses with glass. *Proceedings of the SPIE*, **4948**, 393–400.

142 Watanabe, W. and Itoh, K. (2001) Spatial coherence of supercontinuum emitted from multiple filaments. *Japanese Journal of Applied Physics*, **40**, 592–595.

143 Horn, A., Khajehnouri, H., Kreutz, E.W., and Poprawe, R. (2002) Time resolved optomechanical investigations on the interaction of laser radiation with glass

in the femtosecond regime. *Proc. LIA, ICALEO 2002*, **94**, 1577–1585.

144 Horn, A. (2003) *Zeitaufgelöste Analyse der Wechselwirkung von ultrakurz gepulster Laserstrahlung mit Dielektrika*. Ph.D. thesis. RWTH-Aachen.

145 Bräuchle, F. (2001) Konversionseffizienz nichtlinearer Kristalle und Halbleiter bei Pulsdauern im Piko- und Femtosekundenbereich. Studienarbeit. RWTH-Aachen.

146 Khajehnouri, H. (2002) Laserinduzierte Prozesse in Gläsern mit zeitaufgelöster Absorptionsspektroskopie und Normarski-Mikroskopie im Femtosekundenbereich. Master's thesis. FH-Emden.

147 Koechner, W. (1996) *Solid-State Laser Engineering*, Vol. 1, Springer Verlag, Berlin, Heidelberg, New York.

148 Horn, A., Kreutz, E.W., and Poprawe, R. (2004) Ultrafast time-resolved photography of femtosecond laser induced modifications in BK7 glass and fused silica. *Applied Physics A: Materials Science & Processing*, **79**(4), 923–925.

149 Horn, A., Kaiser, C., Ritschel, R., Mans, T., Russbüldt, P., Hoffmann, H.D., and Poprawe, R. (2007) Si-K_α-radiation generated by the interaction of femtosecond laser radiation with silicon. *Journal of Physics: Conference Series*, **59**, 159–163.

150 Lindenberg, A.M., Larsson, J., Sokolowski-Tinten, K., Gaffney, K.J., Blome, C., Synnergren, O., Sheppard, J., Caleman, C., MacPhee, A.G., Weinstein, D., *et al.* (2005) Atomic-Scale Visualization of Inertial Dynamics, *Science*, **308**(5720), 392–395.

151 Feurer, T., Morak, A., Uschmann, I., Ziener, C., Schwoerer, H., Förster, E., and Sauerbrey, R. (2001) An incoherent sub-picosecond X-ray source for time-resolved X-ray-diffraction experiments. *Applied Physics B: Lasers and Optics*, **72**(1), 15–20.

152 Mans, T., Russbüldt, P., Kreutz, E.W., Hoffmann, D., and Poprawe, R. (2003) Colquiriite fs-sources for commercial applications. *Proceedings of the SPIE*, **4978**, 38.

153 Mancini, R.C., Shlyaptseva, A.S., Audebert, P., Geindre, J.P., Bastiani, S., Gauthier, J.C., Grillon, G., Mysyrowicz, A., and Antonetti, A. (1996) Stark broadening of satellite lines in silicon plasmas driven by femtosecond laser pulses. *Physical Review E*, **54**(4), 4147–4154.

154 Casnati, E., Tartari, A., and Baraldi, C. (1983) An empirical approach to *k*-shell ionisation cross section by electrons. *Journal of Physics B Atomic and Molecular Physics*, **15**(1), 155–167.

155 Casnati, E., Tartari, A., and Baraldi, C. (1983) An empirical approach to *k*-shell ionisation cross section by electrons (corrigendum). *Journal of Physics B Atomic and Molecular Physics*, **16**(3), 505–505.

156 Gordienko, V.M., Lachko, I.M., Mikheev, P.M., Savel'ev, A.B., Uryupina, D.S., and Volkov, R.V. (2002) Experimental characterization of hot electron production under femtosecond laser plasma interaction at moderate intensities. *Plasma Physics and Controlled Fusion*, **44**(12), 2555–2568.

157 Reich, C., Gibbon, P., Uschmann, I., and Förster, E. (2000) Yield optimization and time structure of femtosecond laser plasma K_α Sources. *Physical Review Letters*, **84**(21), 4846–4849.

158 Mahlmann, D., Jahnke, J., and Loosen, P. (2008) Rapid determination of the dry weight of single, living cyanobacterial cells using the Mach–Zehnder double-beam interference microscope. *European Journal of Phycology*, **43**(4), 355–364.

159 Walter, F. (1963) Interferenzmikroskopie und hämatologische Forschung. *Annals of Hematology*, **9**(5), 297–314.

160 Fomin, N.A. (1998) *Speckle Photography for Fluid Mechanics Mesurements*, Springer Verlag, Berlin, Heidelberg.

161 Russbüldt, P. (1996) Abtrag von Metallen und Halbleitern bei der Bearbeitung mit intensiven, ultrakurzen Laserpulsen. Master's thesis. RWTH-Aachen.

162 Allen, R.D., David, G.B., and Nomarski, G. (1969) The Zeiss–Nomarski differential interference equipment for transmitted-light microsospy. *Zeitschrift für Wissenschaftliche Mikroskopie und Mikroskopische Technik*, **69**(4), 193–221.

163 Preza, C., Snyder, D.L., and Conchello, J.-A. (1999) Theoretical development and experimental evaluation of imaging models for differential interference contrast

microscopy. *Journal of the Optical Society of America A*, **16**(9), 2185–2199.

164 Horn, A., Weichenhain, R., Kreutz, E.W., and Poprawe, R. (2001) Dynamics of laser-induced cracking in glasses at a picosecond time scale. *Proceedings of the SPIE*, **4184**, 539–544.

165 Horn, A., Mingareev, I., and Miyamoto, I. (2006) Ultra-fast diagnostics of laser-induced melting of matter. *JLMN-Journal of Laser Micro/Nanoengineering*, **1**(3), 264–268.

166 Wawers, W. (2008) *Präzisions-Wendelbohren mit Laserstrahlung*. Ph.D. thesis. RWTH-Aachen.

167 Walther, K. (2008) *Herstellung von Formbohrungen mit Laserstrahlung*. Ph.D. thesis. RWTH-Aachen, in preparation.

168 König, J., Nolte, S., and Tünnermann, A. (2005) Plasma evolution during metal ablation with ultrashort laser pulses. *Optics Express*, **13**(26), 10597–10607.

169 Bäuerle, D. (1996) *Laser Processing and Chemistry*, Springer Verlag, Berlin.

170 Kelly, R. and Miotello, A. (1999) Contribution of vaporization and boiling to thermal-spike sputtering by ions or laser pulses. *Physical Review E*, **60**(3), 2616–2625.

171 Mingareev, I. and Horn, A. (2007) Time-resolved investigations of plasma and melt ejections in metals by pump–probe shadowgraphy. *Applied Physics A*, **92**(4), 917–920.

172 Korte, F., Koch, J., and Chichkov, B.N. (2004) Formation of microbumps and nanojets on gold targets by femtosecond laser pulses. *Applied Physics A: Materials Science & Processing*, **79**(4), 879–881.

173 Jandeleit, J., Urbasch, G., Hoffmann, H.D., Treusch, H.G., and Kreutz, E.W. (1996) Picosecond laser ablation of thin copper films. *Applied Physics A: Materials Science & Processing*, **63**(2), 117–121.

174 Jandeleit, J., Russbüldt, P., Treusch, H.G., and Kreutz, E.W. (1997) Micromachining by picosecond laser radiation: fundamentals and applications. *Proceedings of the SPIE*, **3097**, 252.

175 Zel'dovich, Y.B. and Raizer, P. (1966) *Physics of shock waves and high-temperature* *hydrodynamic phenomena*, Vol. 1, Academic Press.

176 Jandeleit, J., Russbüldt, P., Urbasch, G., Hoffmann, D., Treusch, H.-G., and Kreutz, E.W. (1996) Investigation of laser-induced ablation processes and production of microstructures by picosecond laser pulses. ICACEO'96, Laser Material Processing, Detroit (USA), 14–17 October 1996, E89–E91.

177 Talkenberg, M., Kreutz, E.W., Horn, A., Jacquorie, M., and Poprawe, R. (2003) UV laser radiation-induced modifications and microstructuring of glass. *Proceedings of the SPIE*, **4637**, 258.

178 Ligbado, G., Horn, A., Kreutz, E.W., Krauss, M.M., Siedow, N., and Hensel, H. (2005) Coloured marking inside glass by laser radiation. *Proceedings of the SPIE*, **5989**, 59890K.

179 Alfano, R.R. (1989) *Supercontinuum Laser Sources*, Springer Verlag, Berlin, Heidelberg.

180 Alfano, R.R. and Shapiro, S.L. (1970) Observation of self-phase modulation and small-scale filaments in crystals and glasses. *Physical Review Letters*, **24**, 592–594.

181 Martin, P., Guizard, S., Daguzan, P., Petite, G., D'Oliveira, P., Meynadier, P., and Perdrix, M. (1997) Subpicosecond study of carrier trapping dynamics in wide-band-gap crystals. *Physical Review B*, **55**, 5799–5810.

182 Quere, F., Guizard, S., Martin, P., Petite, G., Gobert, O., Meynadier, P., and Perdrix, M. (1998) Subpicosecond studies of carrier dynamics in laser induced breakdown. *Proceedings of the SPIE*, **3578**, 10–19.

183 Saeta, P.N. and Greene, B.I. (1993) Primary relaxation processes at the band edge of SiO_2. *Physical Review Letters*, **70**, 3588–3591.

184 Joosen, W. *et al.* (1992) Femtosecond multiphoton generation of the self-trapped exciton in α-SiO_2. *Applied Physics Letters*, **61**(19), 2260–2262.

185 Guizard, S. and Meynardier, M. (1996) Time-resolved study of laser-induced colour centres in SiO_2. *Journal of Physics: Condensed Matter*, **8**, 1281–1290.

186 Petite, G., Daguzan, P., Guizard, S., and Martin, P. (1997) Ultrafast processes in laser irradiated wide bandgap insulators. *Applied Surface Science*, **109/110**, 36–42.

187 Wortmann, D., Ramme, M., and Gottmann, J. (2007) Refractive index modification using fs-laser double pulses. *Optics Express*, **15**(16), 10149–10153.

188 Schaffer, C.B., García, J.F., and Mazur, E. (2003) Bulk heating of transparent materials using a high-repetition-rate femtosecond laser. *Applied Physics A: Materials Science & Processing*, **76**(3), 351–354.

189 Miyamoto, I., Horn, A., and Gottmann, J. (2007) Local melting of glass material and its application to directfusion welding by ps-laser pulses. *JLMN-Journal of Laser Micro/Nanoengineering*, **2**(1), 7–14.

190 Miyamoto, I., Horn, A., Gottmann, J., Wortmann, D., and Yoshino, F. (2007) High-precision, high-throughput fusion welding of glass using femtosecond laser pulses,. *JLMN-Journal of Laser Micro/Nanoengineering*, **2**(1), 57–63.

191 Tamaki, T., Watanabe, W., Nishii, J., and Itoh, K. (2005) Welding of transparent materials using femtosecond laser pulses. *Japanese Journal of Applied Physics*, **44**(22), L687–L689.

192 Watanabe, W., Onda, S., Tamaki, T., Itoh, K., and Nishii, J. (2006) Space-selective laser joining of dissimilar transparent materials using femtosecond laser pulses. *Applied Physics Letters*, **89**, 021106.

193 Tamaki, T., Watanabe, W., and Itoh, K. (2006) Laser micro-welding of transparent materials by a localized heat accumulation effect using a femtosecond fiber laser at 1558 nm. *Optics Express*, **14**(22), 10460–10468.

194 Watanabe, W., Onda, S., Tamaki, T., and Itoh, K. (2007) Direct joining of glass substrates by 1 kHz femtosecond laser pulses. *Applied Physics B: Lasers and Optics*, **87**(1), 85–89.

195 Horn, A., Mingareev, I., Werth, A., Kachel, M., and Brenk, U. (2008) Investigations on ultrafast welding of glass-glass and glass-silicon. *Applied Physics A*, **93**(1), 171–175.

196 Horn, A., Mingareev, I., and Werth, A. (2008) Investigations on melting and welding of glass by ultra-short laser

Radiation. *JLMN-Journal of Laser Micro/Nanoengineering*, **3**(2), 114–118.

197 Kern, W. (1990) The evolution of silicon wafer cleaning technology. *Journal of The Electrochemical Society*, 137, 1887.

198 Rudd, J.V., Zimdars, D.A., and Warmuth, M.W. (2003) Compact fiber-pigtailed terahertz imaging system. *Proceedings of the SPIE*, **3934**, 27.

199 Thomsen, C., Grahn, H.T., Maris, H.J., and Tauc, J. (1986) Surface generation and detection of phonons by picosecond light pulses. *Physical Review B*, **34**(6), 4129–4138.

200 Röser, F., Rothhard, J., Ortac, B., Liem, A., Schmidt, O., Schreiber, T., Limpert, J., and Tünnermann, A. (2005) 131 W 220 fs fiber laser system. *Optics Letters*, **30**(20), 2754–2756.

201 Klingebiel, S., Röser, F., Ortac, B., Limpert, J., and Tünnermann, A. (2007) Spectral beam combining of Yb-doped fiber lasers with high efficiency. *Journal of the Optical Society of America B*, **24**(8), 1716–1720.

202 Shelby, R.A., Smith, D.R., Nemat-Nasser, S.C., and Schultz, S. (2001) Microwave transmission through a two-dimensional, isotropic, left-handed metamaterial. *Applied Physics Letters*, **78**(4), 489.

203 Caloz, C., Chang, C.C., and Itoh, T. (2001) Full-wave verification of the fundamental properties of left-handed materials in waveguide configurations. *Journal of Applied Physics*, **90**(11), 5483.

204 Busch, K., von Freymann, G., Linden, S., Mingaleev, S.F., Tkeshelashvili, L., and Wegener, M. (2007) Periodic nanostructures for photonics. *Physics Reports*, **444**(3–6), 101–202.

205 Stoian, R. *et al.* (2002) Laser ablation of dielectrics with temporally shaped femtosecond pulses. *Applied Physics Letters*, **80**(3), 353.

206 Bulgakova, N.M., Stoian, R., Rosenfeld, A., Hertel, I.V., and Campbell, E.E.B. (2004) Electronic transport and consequences for material removal in ultrafast pulsed laser ablation of materials. *Physical Review B*, **69**(5), 54102.

207 Iatia Ltd., www.iatia.com.au.

208 Allman, B.E. and Nugent, K.A. (2006) Shape Imaging in Defence Operations. *Land Warfare Conference 2006*, Brisbane.

209 Nugent, K., Paganin, D., and Barty, A. (2006) Phase determination of a radiation wave field, 2 May 2006. US Patent 7,039,553.

210 Landau, L.D. and Lifshitz, E.M. (1987) *Fluid Mechanics*, Pergamon.

211 Mangles, S.P.D., Walton, B.R., Tzoufras, M., Najmudin, Z., Clarke, R.J., Dangor, A.E., Evans, R.G., Fritzler, S., Gopal, A., Hernandez-Gomez, C., Mori, W.B., Rozmus, W., Tatarakis, M., Thomas, A.G.R., Tsung, F.S., Wei, M.S., and Krushelnick, K. (2005) Electron acceleration in cavitated channels formed by a petawatt laser in low-density plasma. *Physical Review Letters*, **94**, 245001–4.

212 Dromey, B., Zepf, M., Gopal, A., Lancaster, K., Wei, M.S., Krushelnick, K., Tatarakis, M., Vakakis, N., Moustaizis, S., Kodama, R., Tampoand, M. Stoeckl, C., Clarke, R., Habara, H., Neely, D., Karsch, S., and Norreys, P. (2006) High harmonic generation in the relativistic limit. *Nature Physics*, **2**, 456–459.

213 Nilson, P.M., Willingale, L., Kaluza, M.C., Kamperidis, C., Minardi, S., Wei, M.S., Fernandes, P., Notley, M., Bandyopadhyay, S., Sherlock, M., Kingham, R.J., Tatarakis, M., Najmudin, Z., Rozmus, W., Evans, R.G., Haines, M.G., Dangor, A.E., and Krushelnick, K. (2006) Magnetic reconnection and plasma dynamics in two-beam laser-solid interactions. *Physical Review Letters*, **97**, 255001–4.

214 Mackinnon, A.J., Patel, P.K., Borghesi, M., Clarke, R.C., Freeman, R.R., Habara, H., Hatchett, S.P., Hey, D., Hicks, D.G., Kar, S., Key, M.H., King, J.A., Lancaster, K., Neely, D., Nikkro, A., Norreys, P.A., Notley, M.M., Phillips, T.W., Romagnani, L., Snavely, R.A., and Stephens, R.B. and Town, R.P.J. (2006) Proton radiography of a laser-driven implosion. *Physical Review Letters*, **97**, 045001–4.

215 Lancaster, K.L., Green, J.S., Hey, D.S., Akli, K.U., Davies, J.R., Clarke, R.J., Freeman, R.R., Habara, H., Key, M.H., Kodama, R., Krushelnick, K., Murphy, C.D., Nakatsutsumi, M., Simpson, P., Stephens, R., Stoeckl, C., Yabu-uchi, T., Zepf, M., and Norreys, P.A. (2007) Measurements of energy transport patterns in solid density laser plasma interactions at intensities of 5×10^{20} W cm^{-2}. *Physical Review Letters*, **98**, 125002–4.

216 Murphy, C.D., Trines, R., Vieira, J., Reitsma, A.J.W., Bingham, R., Collier, J.L., Divall, E.J., Foster, P.S., Hooker, C.J., Langley, A.J., Norreys, P.A., Fonseca, R.A., Fiuza, F., Silva, L.O., Mendonca, J.T., Mori, W.B., Gallacher, J.G., Viskup, R., Jaroszynski, D.A., Mangles, S.P.D., Thomas, A.G.R., Krushelnick, K., and Najmudin, Z. (2006) Evidence of photon acceleration by laser wake fields. *Physics of Plasmas*, **13**, 033108–8.

217 Kodama, R., Shiraga, H., Shigemori, K., Toyama, Y., Fujioka, S., Azechi, H., Fujita, H., Habara, H., Hall, T., Izawa, Y., *et al.* (2002) Nuclear fusion: Fast heating scalable to laser fusion ignition. *Nature*, **418**(6901), 933–934.

218 Edwards, G.S., Allen, S.J., Haglund, R.F., Nemanich, R.J., Redlich, B., Simon, J.D., and Yang, W.-C. (2005) Applications of free-electron lasers in the biological and material sciences. *Photochemistry and Photobiology*, **81**, 711–735, doi:10.1111/j.1751-1097.2005.tb01437.x.

Glossary

Amplifier Generally, an amplifier is any device that changes, usually increases, the amplitude of a signal. The "signal" is usually voltage, current, or optical radiation. An ultra-fast laser amplifier increases the power to several 100 W or pulse energies > 1 mJ of an ultra-fast laser oscillator if possible by not changing the other radiation properties.

Autocorrelation Autocorrelation is used for finding repeating patterns, such as the presence of a periodic signal which has been buried under noise, or identifying the missing fundamental frequency in a signal implied by its harmonic frequencies. It is used frequently in signal processing for analyzing functions or series of values, such as time domain signals, like laser pulses. Autocorrelation is used in ultra-fast laser technology for the measurement of the pulse duration.

BBP The beam parameter product is the product of a laser beam's divergence angle (half-angle) and the radius of the beam at the beam waist. The BPP quantifies the quality of the laser radiation, and how well it can be focused to a small spot.

Caustic In optics, a caustic is the envelope of light rays reflected or refracted by a curved surface or object, or the projection of that envelope of rays on another surface. Caustic refers in laser technology to the curve to which light rays are tangent, defining a boundary of an envelope of rays as a curve of concentrated light, in other words, the laser focus.

Chirp A chirp is a signal at which the frequency increases ("up-chirp") or decreases ("down-chirp") with time. It is commonly used in sonar and radar, but has other applications, such as in spread spectrum communications. In optics, ultra-fast laser radiation also exhibits chirp due to the dispersion of the materials they propagate through.

Coherence Coherence is a property of waves, that enables stationary (in other words, temporally and spatially constant) interference. More generally, coherence describes all correlation properties between physical quantities of a wave.

Correlation Correlation (often measured as a correlation coefficient) indicates the strength and direction of a linear relationship between two random variables.

CPA Chirped pulse amplification (CPA) is a technique for amplifying an ultra-short laser pulse up to the petawatt level with the laser pulse being stretched out temporally and spectrally prior to amplification. CPA is the current state of

Ultra-fast Material Metrology. Alexander Horn
Copyright © 2009 WILEY-VCH Verlag GmbH & Co. KGaA, Weinheim
ISBN: 978-3-527-40887-0

the art technique to which all of the highest power lasers (greater than about 100 terawatts) are measured.

Cross-section In nuclear and particle physics, the concept of a cross-section is used to express the likelihood of interaction between particles. When particles are thrown against a foil made of a certain substance, the cross-section describes a hypothetical area measured around the target particles (usually its atoms) that represents a surface. If a particle of the beam crosses this surface, there will be some kind of interaction.

Dielectric A dielectric is a nonconducting substance, in other words, an insulator, like glass, ceramics and gases.

Dispersion Dispersion is the phenomenon in which the phase velocity of a wave depends on its frequency. Media having such a property are termed dispersive media.

FEL A free-electron laser, or FEL, is a laser that shares the same optical properties as conventional lasers such as emitting optical radiation consisting of coherent electromagnetic radiation, but which uses some very different operating principles to form the beam. Unlike gas, liquid, or solid-state lasers such as diode lasers, in which electrons are excited in bound atomic or molecular states, FELs use a relativistic electron beam as the lasing medium which moves freely through a magnetic structure, hence the term free electron.

Gaussian beam In optics, a Gaussian beam is a beam of electromagnetic radiation whose transverse electric field and intensity distributions are described by Gaussian functions. Many laser sources emit radiation with a Gaussian spatial intensity profile. The laser is operating on the fundamental transverse "TEM_{00} mode" of the laser's optical resonator.

Group velocity The group velocity of a wave is the velocity with which the overall shape of the wave's amplitudes (known as the modulation or envelope of the wave) propagate through space. The definition of group velocity is only useful for wave packets, which is a pulse that is localized in both real space and frequency space. Because waves at different frequencies propagate at differing phase velocities in dispersive media, for a large frequency range (a narrow envelope in space) the observed pulse would change shape while traveling, making group velocity an unclear or useless quantity.

GVD Group velocity dispersion: The group velocity itself is usually a function of the wave's frequency. This results in group velocity dispersion, which causes a short pulse of light to spread in time as a result of different frequency components of the pulse traveling at different velocities.

Kerr effect Also the quadratic electro-optic effect (QEO effect), is a change in the refractive index of a material in response to an electric field. It is distinct from the Pockels effect in that the induced index change is directly proportional to the square of the electric field instead of to the magnitude of the field. All materials show a Kerr effect, but certain liquids display the effect more strongly than other materials do.

Laser Light amplification by stimulated emission of radiation. A non-natural physical process transforming energy, like electricity, heat, or incoherent radiation with bad quality into optical energy with very good properties, like coherence, focusable, monochromatic, ultrashort pulsed radiation. A laser consists generally of a laser-active solid, liquid or gaseous medium, two mirrors, one being highly reflective the other partially reflective, and a pumping system stimulating the laser-active medium.

LTE Local thermodynamic equilibrium: It is useful to distinguish between global and local thermodynamic equilibrium. In thermodynamics, exchanges within a system and between the system and the outside are controlled by intensive parameters. As an example, temperature controls heat exchanges. Global thermodynamic equilibrium (GTE) means that those intensive parameters are homogeneous throughout the whole system, while local thermodynamic equilibrium (LTE) means that those intensive parameters are varying in space and time, but are varying so slowly that for any point, one can assume thermodynamic equilibrium in some neighborhood about that point.

Mode-locking Mode-locking is a technique in optics by which a laser can be made to produce pulses of light of extremely short duration, on the order of picoseconds (10^{-12} s) or femtoseconds (10^{-15} s).

MOPA The most popular way of achieving power scalability is the "MOPA" (master oscillator/power amplifier) approach. The master oscillator produces a highly coherent beam, and an optical amplifier is used to increase the power of the beam while preserving its main properties. The master oscillator has no need to be powerful, and has no need to operate at high efficiency because the efficiency is determined mainly by the power amplifier. The combination of several laser amplifiers seeded by a common master oscillator is essential concept of the High Power Laser Energy Research Facility.

Nonlinear optics Nonlinear optics (NLO) is the branch of optics that describes the behavior of light in nonlinear media. This media's dielectric polarization P responds nonlinearly to the electric field E of the radiation. This nonlinearity is typically observed at very large radiation intensities, such as those provided by ultra-fast lasers.

Oscillator Oscillation is the repetitive variation, typically in time, of some measure about a central value (often a point of equilibrium) or between two or more different states. An oscillator is given in a laser by the mirrors, redirecting the amplified radiation back into the laser-active medium. Ultra-fast laser oscillators represent radiation sources exhibiting a low output power < 10 W and a large repetition rate > 1 MHz.

Phase velocity The phase velocity (or phase speed) of a wave is the rate at which the phase of the wave propagates in space and is the speed at which the phase of any one frequency component of the wave travels. For such a component, any given phase of the wave (for example, the crest) will appear to travel at the phase velocity.

Pockels effect Also Pockels electro-optic effect, produces birefringence in an optical medium induced by a constant or varying electric field. It is distinguished from the Kerr effect by the fact that the birefringence is proportional to the electric field, whereas in the Kerr effect it is quadratic in the field. The Pockels effect occurs only in crystals that lack inversion symmetry, such as lithium niobate or gallium arsenide and in other non-centrosymmetric media such as electric-field poled polymers or glasses.

PSF The point spread function (PSF) describes the response of an imaging system to a point source or point object.

SHG Second harmonic generation (also frequency doubling) is a nonlinear optical process, in which photons interacting with a nonlinear material are locked together resulting in twice the energy, and therefore twice the frequency and half the wavelength of the initial photons.

SPM Self-phase modulation is a nonlinear optical effect of radiation-matter interaction. An ultrashort optical pulse passing through a medium will induce a varying refractive index of the medium due to the optical Kerr effect. This variation in refractive index will produce a phase shift in the pulse, leading to a change of the pulse's frequency spectrum.

Ultra-fast laser spectroscopy Ultra-fast laser spectroscopy is the study of molecules on extremely short time scales (nanoseconds to femtoseconds) after their excitation with a pulsed laser.

Ultrashort pulse An ultrashort pulse of light is an electromagnetic pulse whose time duration is on the order of the femtosecond (10^{-15} s). Such pulses have a broadband optical spectrum, and can be created by mode-locked oscillators.

Wave packet In physics, a wave packet is an envelope or packet containing a number of plane waves having different wavelengths, chosen such that their phases and amplitudes interfere constructively over a small region of space. Depending on the evolution equation, while propagating the wave packet's envelope may remain constant (no dispersion) or it may change (dispersion). In this way a fixed phase relationship between the modes of the laser's resonant cavity is induced. The laser is then phase-locked or mode-locked. Interference between these modes causes the laser radiation to be generated as a train of pulses. Depending on the properties of the laser, these pulses may be of extremely small duration, as short as a few femtoseconds. They are characterized by a high peak intensity that usually leads to nonlinear processes in various materials. These processes are studied in the field of nonlinear optics.

Index

Ultra-fast Material Metrology. Alexander Horn
Copyright © 2009 WILEY-VCH Verlag GmbH & Co. KGaA, Weinheim
ISBN: 978-3-527-40887-0